STAR TREK®

THE OFFICIAL GUIDE TO OUR UNIVERSE

THE OFFICIAL GUIDE TO OUR UNIVERSE

THE TRUE SCIENCE BEHIND THE STARSHIP VOYAGES

Andrew Fazekas

Foreword by William Shatner

NATIONAL GEOGRAPHIC

WASHINGTON, D.C.

CONTENTS

PREVIOUS PAGES: The U.S.S. Enterprise *flees an android-inhabited planet in the original series (TOS / "I, Mudd").*

OPPOSITE: Spock, Captain James T. Kirk, Leonard H. "Bones" McCoy, and Montgomery "Scotty" Scott find themselves encircled by a force field at the O.K. Corral (TOS / "Spectre of the Gun").

FOREWORD

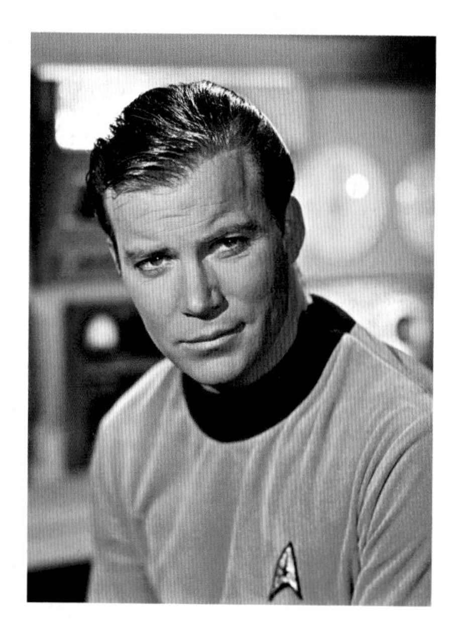

Talosian mirages lure Captain Christopher Pike and his crew to Talos IV in Star Trek's *first pilot, "The Cage."*

Left: William Shatner as Captain Kirk in TOS

Science fiction is essentially mythological. Like religion, it seeks answers to questions of depth: Who am I? Where am I going? What is the meaning of my life? It is the imagination of talented writers seeking answers to other unknowables: Are there UFOs? Does life in our complex form exist in the universe? Is there a higher intelligence? Is there any way to span the vast distances not only in the universe (or maybe parallel universes) but between molecules and atoms and nuclei and quarks and bosons?

Out of the fertile imaginations of the great science fiction writers, we become intrigued and mesmerized by the possibilities. We are fascinated by the tantalizing look we get when science opens a window and we peek for an instant into a mysterious realm that is both forbidding and intriguing. What's out there? Who's out there and why? Science fiction dares to answer these questions—and *Star Trek,* over the years, has done just that.

The majesty of what we can see and feel and hear and touch is extraordinary. The vastness of the unanswerable view of quantum physics is dazzling and sometimes defeating. Science fictions, like the epic stories told about the *Star Trek* universe, give us a taste of what it might be like

to truly grasp the yet unknowable. That is why this book, which shows the connections between today's science and *Star Trek*'s tomorrows, may summon up new questions for you.

Questions, questions, questions, ripping and tearing, banging and drilling in our minds, and no answers, or maybe one or two. It's possible that in this book, you may find not only a lot of important information but also, just maybe, a question that is yet to be asked. Don't you think?

— *William Shatner*

INTRODUCTION

THE STAR TREK *UNIVERSE, WHICH NOW CELEBRATES* 50 years, is a pop culture phenomenon that transcends science fiction and is beloved across the globe. Since that very first episode beamed into our living rooms on September 8, 1966, it has continued to capture the imagination of generations of fans, spawning six TV series, numerous documentaries, countless fan-made films, and 13 feature movies—and counting. *Star Trek* attracts followers thanks to its hallmark blend of thrilling quests to the far reaches of the galaxy, a hopeful view of Humanity's future, and ringside seats to some of the greatest cosmic wonders of nature.

Like any good tale of the fantastic, *Star Trek* has always focused on discovering and showcasing places beyond the audience's immediate environment—in other words, exploring strange new worlds. From its birth, this groundbreaking franchise has taken the science of astronomy seriously and has always drawn on contemporary cosmological discoveries. Although *Star Trek*'s creators have been off the mark scientifically in a few instances—especially in the original series, when at the time our knowledge of the cosmos and its clockwork were vastly more limited than today—it's amazing how much they did get right.

For its creator, Gene Roddenberry, and many of *Star Trek*'s other writers, the vision for this grand space opera has always been about great storytelling accomplished, no doubt, by taking artistic license, but the *Star Trek* adventure has also always embraced real-world science and technology pushed to their boundaries to create a universe in which Humans can explore the stars. With science consultants on board, *Star Trek* voyages have always played out on a stunning interstellar canvas filled with exotic sights like colorful nebulae, alien worlds, and supernovae, just to name a few. What those voyages envisioned, our most advanced technologies are now exploring for real. Recent findings in astronomy and astrophysics today are making our own universe as fun, exciting, and unpredictable as the universe we love to travel in *Star Trek*.

On a more personal note, this project is like a Vulcan mind-meld of sorts, with two of my passions fusing into this one tome. Some of my earliest childhood memories are from the early 1970s, watching reruns of the original *Star Trek* series with my father. I was immediately hooked, mesmerized by the whole idea of the vastness of space. While I found its enormity a bit chilling, I couldn't help but admire how Captain Kirk and his crew were so daring, brave, and inquisitive. It made me want to explore space and make those connections myself. Sitting on my father's lap and viewing the heavens through our telescope from the rooftop of our apartment building in Montreal allowed me to do just that.

This fascination with both stargazing and *Star Trek* has followed me into adulthood, and now I find I'm passing my passions on to my own two young daughters, who not only are just discovering the voyages of

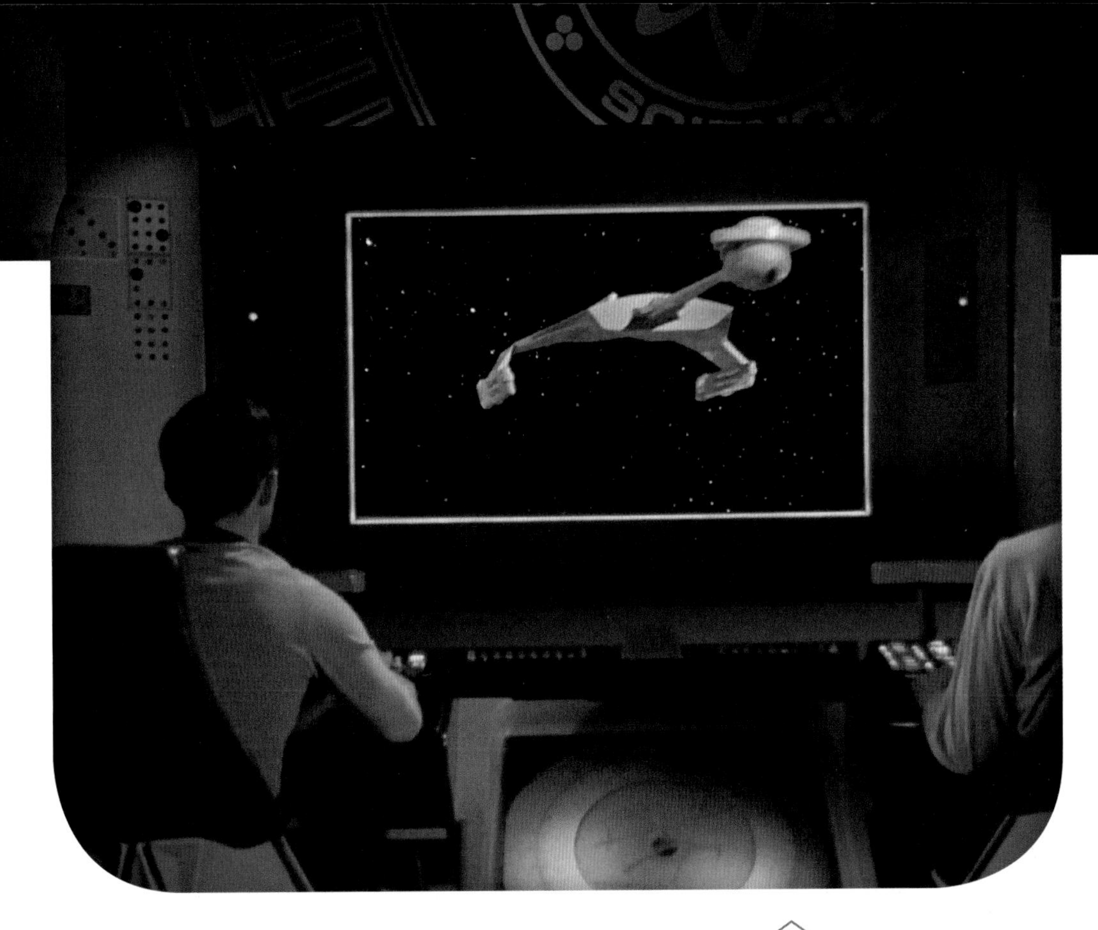

A Klingon warship initially mistaken for a sensor "ghost" tails the U.S.S. Enterprise *as it transports Elaan of the ruling Elasian dynasty to the enemy Troyian planet for an arranged marriage that's hoped to end the interplanetary war (TOS / "Elaan of Troyius").*

the *Enterprise,* tribbles, and the holodeck for themselves, but also are embarking on their own journey of exploring science and the night sky filled with constellations and their treasures.

As William Shatner states in his foreword, science can open a window to the mysterious realm of the universe. And for its fandom, *Star Trek*'s adventure tales offer a chance to peer into that vast cosmic landscape and let our imaginations soar.

So I invite you to join me in connecting the *Star Trek* universe with our own, and let's boldly go where no one has gone before.

— *Andrew S. Fazekas, Stardate 2016.115*

TREKKING THE NIGHT SKY

STAR TREK *STARSHIPS HAVE TAKEN US TO MANY* fantastical star systems and far-flung worlds, most within the realm of our own galaxy, the Milky Way, yet all implying a much larger universe, one that is beyond the reach of even the fastest ship in Starfleet. The distances are incredible, even by *Star Trek* scale: A starship traveling at warp factor nine would take more than six decades to travel across one-tenth of the Milky Way and well over a millennium to reach the nearest neighboring galaxy. This unfathomably vast cosmos is the stage on which all the *Star Trek* voyages take place, inviting us to escape our homeworld and see the universe from a whole new perspective.

THE SCALE OF THE UNIVERSE

Using a series of ever widening astronomical steps, we can begin to grasp the overwhelming size of the universe. We begin with our Earth, a small rocky world with a single moon. Take one step outward, and we find that Earth is just one of a family of eight major planets and thousands of smaller bodies orbiting a star they share—the sun. Known collectively as the solar system, this family of planets and other bits of matter is amazingly large to begin with. Dwarf planets like Pluto and thousands of sibling icy bodies that make up the disk-like Kuiper belt exist as far out as about 8 billion miles (13 billion km) from the sun.

Take another step back and we find that our sun, a common yellow dwarf star, is just one speck of a star among 400 billion others that form the Milky Way, the visible portion of which stretches 100,000 light-years across, with an additional halo of dark matter approximately 400,000 light-years across.

A short digression to talk about the relationship between time and space here. Distances between the stars and galaxies are so great that conventional yardsticks such as miles or kilometers become impractical. So to measure distances beyond the solar system, astronomers use the unit of the light-year: the distance light travels in an Earth year. Light travels at a speed of about 180,000 miles (300,000 km) per second, and so one light-year is equal to about 6 trillion miles (10 trillion km). Because these distances represent the speed of light traveling from afar, when astronomers use powerful telescopes to peer far off into the universe, they are also peering far back into time. In fact, astronomers' instruments today can see so far away, they are looking as far back as 13.8 billion years in time and glimpsing some of the universe's first stars and galaxies being born, within a half billion years after the birth of the universe itself, known as the big bang. And imagine, that is just the universe we can see!

Even the Milky Way holds mysteries, both within the *Star Trek* world and our own.

OUR HOME GALAXY

The home galaxy of Earth, the United Federation of Planets, Vulcans, Klingons, and nearly all alien species encountered in the *Star Trek* universe, is the Milky Way. It is shaped like a flattened pinwheel, and our sun is tucked away in the suburbs of the galaxy, in one of its outer spiral arms. In fact, today's astronomers have only a basic sense of the galaxy's overall spiral structure. We know it has four major spiral arms, with two of them more prominent, wrapping around what is likely a bar-shaped nucleus. But there is still much unknown, because our views are still closely tied to planet Earth.

In *Star Trek,* stellar cartographers knew enough about the Milky Way to subdivide it into four quadrants—Alpha, Beta, Gamma, and Delta—even though Humans had charted less than a fifth of the galaxy by

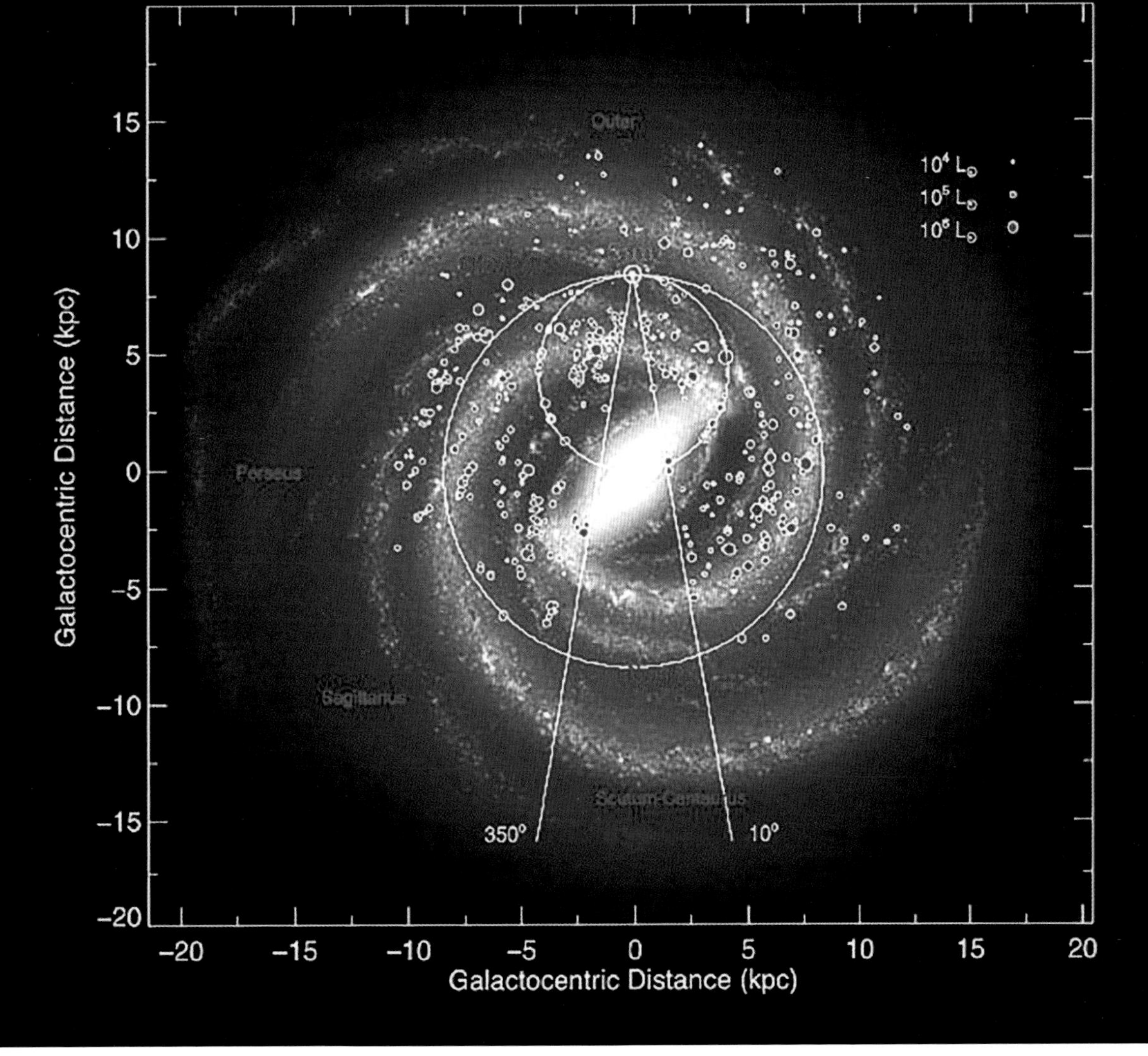

the 24th century. At the galaxy's heart, so the *Star Trek* cartographers believed, was a core where matter was created, surrounded by an impenetrable energy force called the Great Barrier. Here was home to the planet Sha Ka Ree, the Vulcan equivalent of Eden. Our astronomers see it differently, theorizing that at the core of the Milky Way hides a supermassive black hole known as Sagittarius A*, billions of times more massive than our sun.

THE JOURNEY CONTINUES

With mystery at its core, our Milky Way appears to belong to a loose association of about 50 or so galaxies

Plotting young massive stars (blue circles) and H II regions (red circles)—often the birthplace of stars—helps confirm the Milky Way's shape as a four-armed spiral. Galactic distances are measured in kiloparsecs (kpc): 1 kpc equals 3,262 light-years.

that form the Local Group, which spreads about 10 million light-years across and nestles in the outskirts of a dense cloud of 2,000 or more galaxies great and small.

As we journey outward, we travel in our minds, guided by science, to realms that lie almost beyond our human imagining. We have a lot to learn and experience—and *Star Trek* leads the way.

ABOUT THIS BOOK

I*N THIS BOOK WE PULL TOGETHER A VAST COLLECTION* of some of the most fascinating astronomical objects visited by the famous starships and iconic characters from throughout the *Star Trek* films and television shows. We first introduce these celestial phenomena through the familiar lens of *Star Trek*, looking at the role they play in the *Trek* stories told—past, present, and future. We then investigate the real science behind them, exploring the counterpoint between the imagined *Star Trek* universe and the facts of our solar system, galaxy, and beyond.

LIVE THE STAR TREK *MOMENTS*

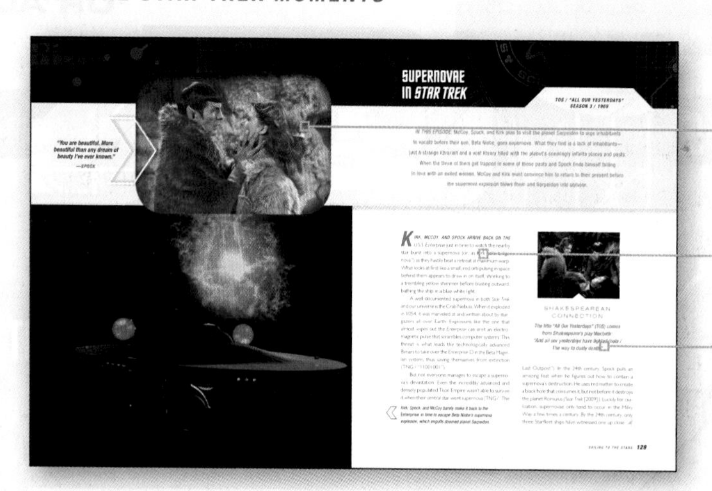

EPISODE HIGHLIGHTS *Experience Starfleet's encounter with real-life space objects, as they play out in a featured episode and beyond.*

LOOK CLOSER *Investigate exciting space topics in the* Star Trek *universe by way of episode and movie scenes.* *See pp. 229-232 for a complete list of episodes referenced.*

GET IN THE SCENE *Peer deeper into the* Star Trek *universe with interesting bits of trivia about the featured episode, space object, and more.*

DISCOVER THE REAL-LIFE SCIENCE

UNDERSTANDING OUR UNIVERSE *Explore the fascinating true science of the space objects and systems that make up our universe.*

THROUGH THE TELESCOPE *Glimpse the far reaches of our galaxy via images captured by spacecrafts or by high-powered telescopes stationed on Earth.*

ARE WE THERE YET? *From tractor beams to time travel, compare* Star Trek's *visionary technology to what we have—or might someday have—in real life.*

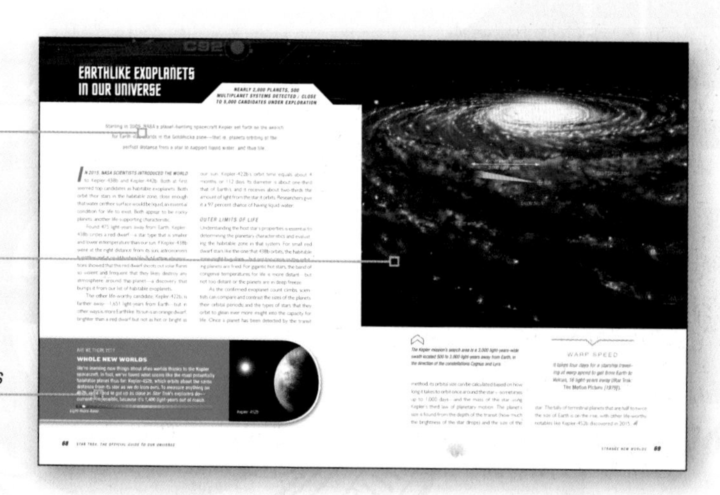

Finally, we guide you through night sky observations, so you can learn to observe a variety of natural space objects that have played pivotal roles and served as dramatic backdrops in your favorite *Trek* scenes. With the help of straightforward instructions, you'll be able to step outside, navigate the night sky, and glimpse them with your own eyes, binoculars, and small backyard telescopes.

SEARCH THE NIGHT SKY

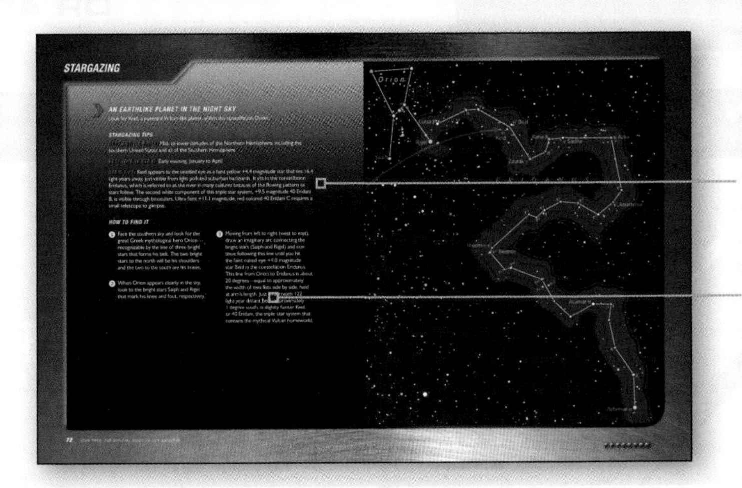

VISUAL GUIDANCE *Use star maps to search for constellations and "star hop" across the sky.*

1, 2, 3 STEPS *Observe the space objects portrayed in* Star Trek *from your backyard with simple instructions oriented for northern latitude locations.*

KEY TO THE STAR CHARTS

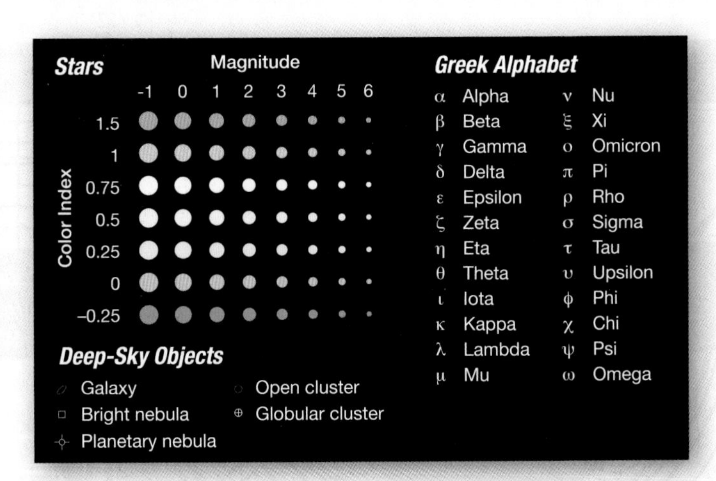

The magnitude of stars shown in the star maps is indicated by size; the color, primarily driven by surface temperature, is indicated as well. Special symbols stand for deep-sky objects, as shown. It can also help to know the Greek alphabet, often used to name stars in a constellation.

NCC-1701-D

THE TERRAN SYSTEM

THERE'S PLENTY TO EXPLORE AMONG THE PLANETS ORBITING SOL, OUR HOMEWORLD STAR.

CHAPTER ONE

STAR TREK & US

No matter how broad the known universe, no matter how numerous the planets and star systems explored, Earth (or Terra) remains in *Star Trek*—as it does for us today—the homeworld. Emerging from the devastation of World War III in the middle of the 21st century, *Star Trek*'s vision of Human history enters a new era on a United Earth. After a decades-long planetary conflict fraught with nuclear attacks and harmful eugenics, deep space exploration comes back into focus. By 2161, United Earth lies at the heart of the United Federation of Planets—a galactic republic allying United Earth with Vulcan, Andoria, Alpha Centauri, and Tellar—that bolsters courageous leaders to venture ever farther into the Milky Way galaxy and beyond. Starfleet's heroic captains are the ones who lead these spacefaring adventures, from Captains Jonathan Archer and James T. Kirk to Jean-Luc Picard, Benjamin Sisko, and Kathryn Janeway. As the Federation's space exploration and defense service, Starfleet's mission is to advance knowledge about the Milky Way and its inhabitants. Yet whatever distance they travel, they remain rooted to the Sol (or Terran) system, which takes its names from the Latin words for "sun" and "earth": *sol* and *terra*.

OUR GALACTIC NEIGHBORHOOD

Star Trek's Sol system is a vast and bustling place. But in the *Star Trek* state of mind, our Sol system—nine planets anchored by the sun—is a tiny piece of a sprawling, busy Milky Way galaxy. The Sol system lies at the heart of the Federation, an organized consortium of more than 150 member worlds stretched across a swath of the galaxy 8,000 to 10,000 light-years in diameter—a distance that takes years for a Starfleet vessel to traverse, even at maximum warp. By the mid-22nd century, space geographers have divided our home galaxy into four quadrants: Alpha, Beta, Gamma, and Delta. Earth and its Sol system reside in the Alpha Quadrant.

In the *Star Trek* universe, Human space exploration starts close to home. As technology progresses, so do the missions—from exploring planets to colonizing them. Humans successfully visit and, in some cases, inhabit planets and moons we know today, from windswept Mars to pockmarked Luna (Earth's moon). *Star Trek: Deep Space Nine* sees the terraforming of Venus, using advanced technology to transform it into a more habitable world despite its harsh weather (DS9 / "Past Tense, Part I"). By the 24th century, our neighboring planets are well traveled enough that a shuttle service called the Jovian Run offers once-a-day travel between Jupiter and Saturn.

As exploration of the real universe continues to push the boundaries of the known space neighborhood outward, perhaps a *Star Trek*–like regional system will arise to

PREVIOUS PAGES: Star Trek: The Next Generation's Enterprise, *the fifth Starfleet vessel to bear that name, explores far corners of the cosmos, but the Sol system will always be its home.*

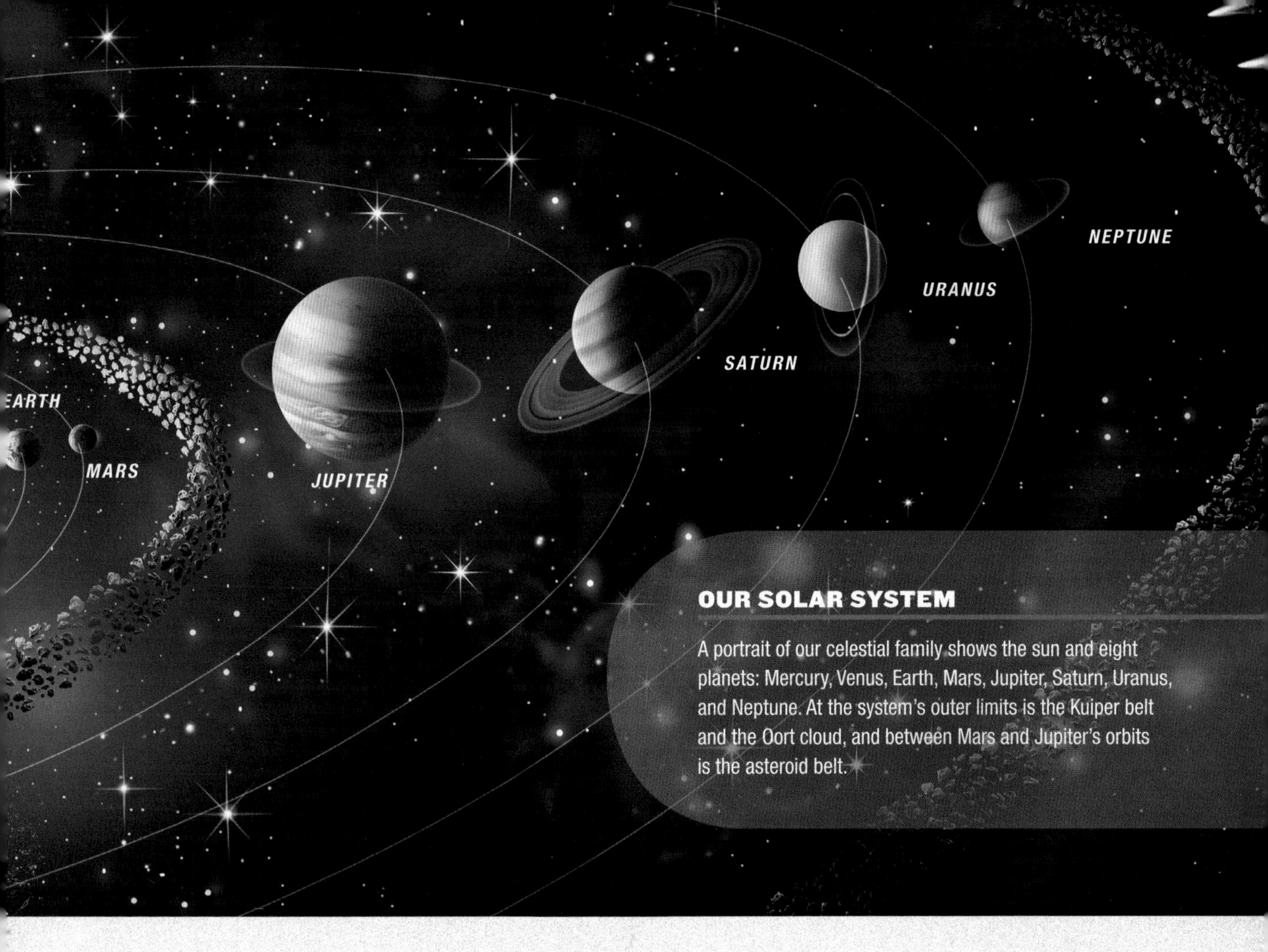

OUR SOLAR SYSTEM

A portrait of our celestial family shows the sun and eight planets: Mercury, Venus, Earth, Mars, Jupiter, Saturn, Uranus, and Neptune. At the system's outer limits is the Kuiper belt and the Oort cloud, and between Mars and Jupiter's orbits is the asteroid belt.

organize the expanding realm. But just as it does in *Star Trek,* our solar system is—and will remain—the cradle of humanity.

WHAT'S IN A SOLAR SYSTEM?

Our solar system was born less than five billion years ago, shielded in a dark and cold cloud of gas and dust and given a shove to collapse by a nearby star explosion. Gravity urged the material to coalesce and condense, causing the cloud to spin faster and then flatten like a pancake. As more and more hydrogen and helium gas were drawn to the nebula's middle, a bulge known as a protostar formed. That protostar would become the sun.

Eventually the surrounding disk started to cool, and ever larger lumps of dust began to form into planetesimals. Over millions of years, these growing bodies continued to collide and consolidate to form the planets we know today, all of which travel along their own slightly elliptical pathways, or orbits, counterclockwise around the sun. Our real-world sun hosts a family of eight major planets—Mercury, Venus, Earth, Mars,

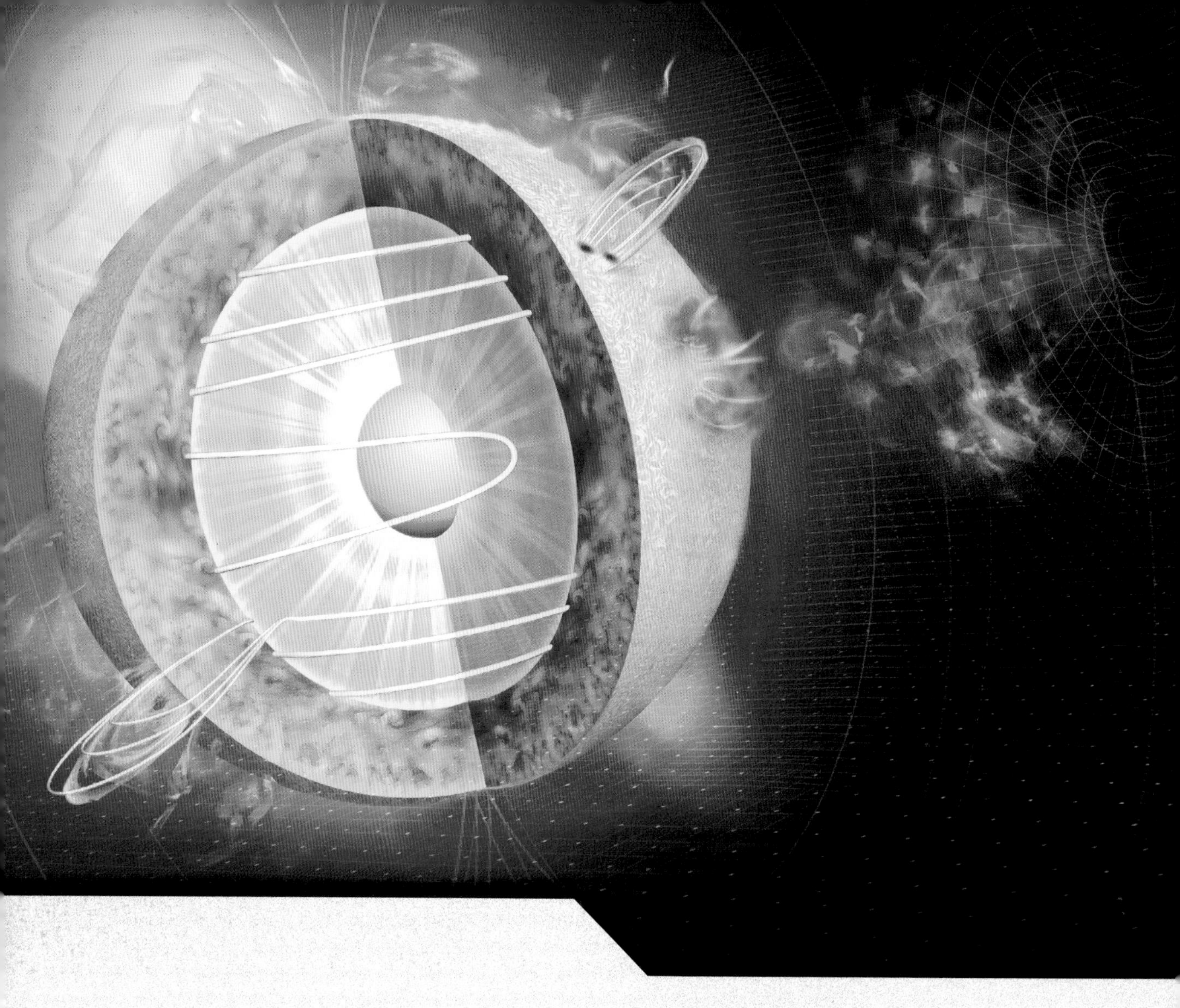

Jupiter, Saturn, Uranus, and Neptune. In the *Star Trek* universe, our solar system consists of nine, because Pluto is considered a planet—that's the kind of alternate history that can arise when comparing the imagined *Star Trek* universe of the future with the scientifically validated universe of the present day.

A planet's relationship to its sun affects its climate and makeup. This principle holds true not only for the planets in our solar system but also for many exoplanets throughout the universe. Likewise, most of the planets envisioned in the *Star Trek* universe take their character from their suns and relationships to them. Terrestrial planets like Mercury, Venus, Earth, and Mars are formed from solid materials because their closeness to our sun made conditions more favorable for rock and metal to condense there. Colder conditions farther from the sun mean that the gas and ice giants—Jupiter, Saturn, Uranus, and Neptune—coalesced and gathered huge envelopes of atmosphere out of lighter, gaseous substances like hydrogen, helium, and water vapor.

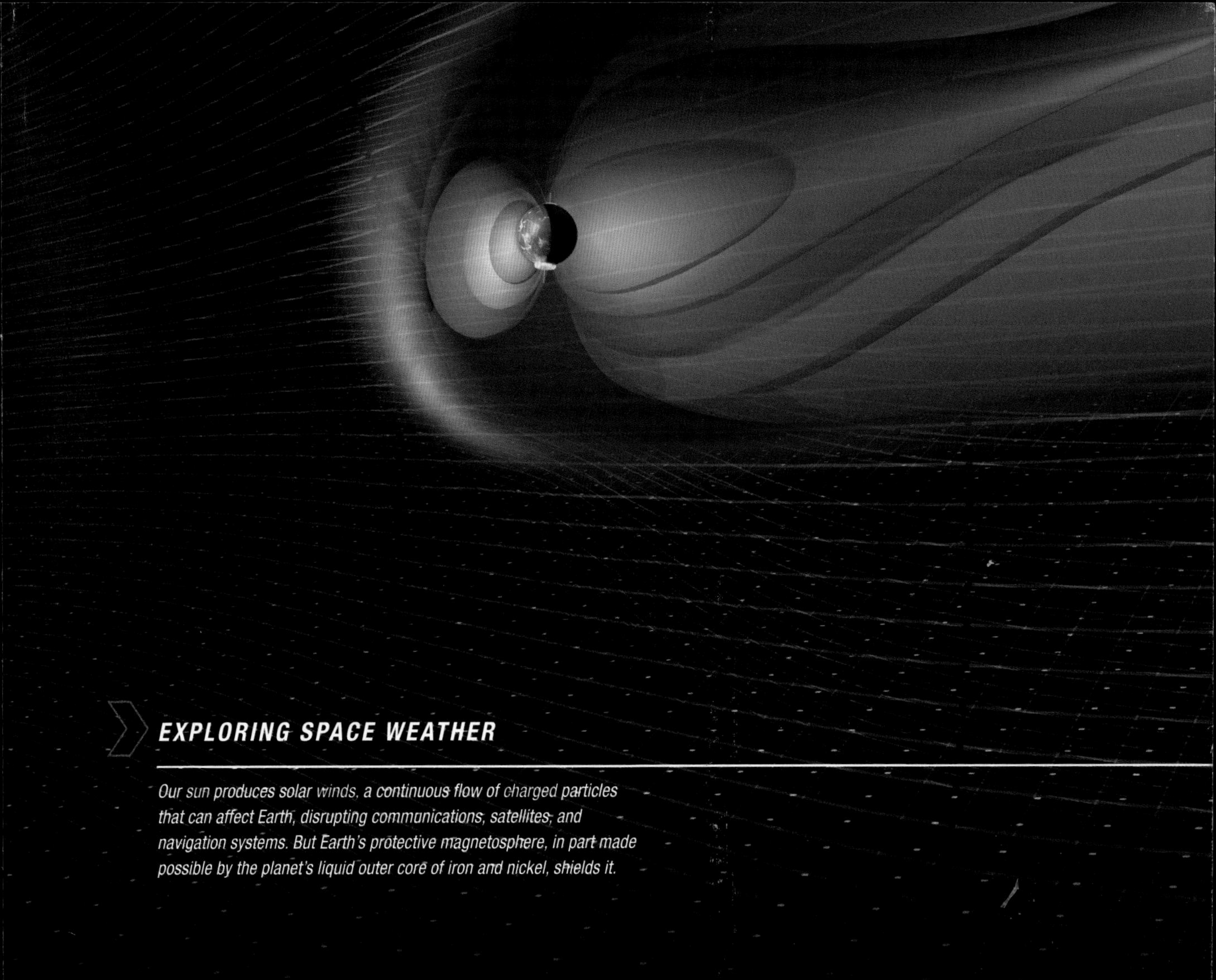

EXPLORING SPACE WEATHER

Our sun produces solar winds, a continuous flow of charged particles that can affect Earth, disrupting communications, satellites, and navigation systems. But Earth's protective magnetosphere, in part made possible by the planet's liquid outer core of iron and nickel, shields it.

OUR SOLAR FURNACE

Our hydrogen- and helium-based sun keeps the solar system alive. Due to the crush of its own gravity, which creates pressure that triggers nuclear reactions at its core, the sun emits energy released as heat and light. Some 600 million tons of hydrogen undergo these reactions every second, and have been for nearly five billion years. Astronomers estimate the sun is only halfway through its life span and it is one of hundreds of billions of stars in our galaxy.

Starships like the *U.S.S. Enterprise* make it possible for researchers in the *Star Trek* universe to visit a myriad of star systems, some dominated by a star not unlike our own sun. Some of these class M planets are quite Earthlike—such as the planet Janeway and Chakotay dub "New Earth" because of how closely it resembles its namesake (VGR / "Resolutions").

Although today's astronomers cannot travel at warp speed, the instruments they have developed reach increasingly farther out, making observations that could lead to the discovery of new planets—and maybe even new life-forms.

"I surrendered to you because, despite your attempt to convince me otherwise, you seem to have a conscience, Mr. Kirk."

—KHAN

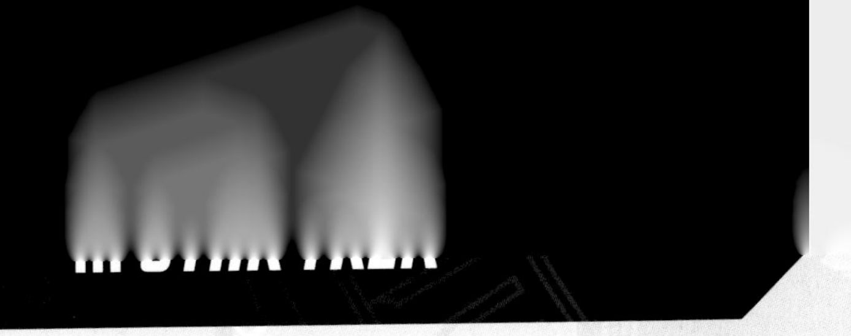

STAR TREK
INTO DARKNESS / 2013

IN THIS MOVIE: After an epic cosmic manhunt, Captain Kirk captures the genetically engineered villain Khan and plans to send him to Earth as a war criminal. Before he can, the *U.S.S. Enterprise* is confronted by the *U.S.S. Vengeance*, commanded by rogue Starfleet Admiral Alexander Marcus, demanding that Khan be handed over. Kirk tries to evade him, but the *Vengeance* opens fire on the *Enterprise*, forcing it to drop out of warp speed near Earth's moon, Luna. Luna provides a dramatic crescent-shaped backdrop for a climactic shoot-out, its crater rims casting long shadows across its uneven surface.

EARTH'S LONE NATURAL SATELLITE IS KNOWN AS Luna in the *Star Trek* universe, its ancient Latin name. Luna is the site of the first permanent Human outpost on a world beyond Earth, founded in the late 21st century. The first solid foothold for Humankind reaching out into the *Star Trek* universe, Luna and its settlements represent an essential jumping-off point for Human expansion into deep space.

Within 300 years of its initial settlement, more than 50 million people are living in the lunar colonies. Lunar Colony One is the largest. Its citizens live under pressure domes, which supply the oxygen and other gases necessary for survival. Outside of the domes, Humans have to wear space suits to maintain an Earthlike atmospheric pressure so they don't explode and take myofibrilin injections to fight the effects of low gravity. The sun rises only once a month in an event called "lunar morning." When she was young, Chief Petty Officer Dorian Collins and her father would hike across the Sea of Clouds to witness the occasion (DS9 / "Valiant").

In the alternate reality depicted in Star Trek Into Darkness *(2013), the* Enterprise *streaks just outside of Luna's orbit, before nearly crashing on Earth as it evades the* Vengeance.

LUNAR BIRTHRIGHT

The U.S.S. Enterprise*-D's Chief Medical Officer Beverly Crusher (TNG) is a Luna native from an early community situated inside prominent moon crater Copernicus.*

Extensive mining operations crop up, collecting minerals that supply the rest of the *Star Trek* universe. The largest of these enterprises, the Orpheus Mining Colony, is designed to be mobile so it can be relocated when it runs out of resources. Humans on Earth develop a derogatory term for Lunar natives: "Lunar schooner," considered an insult by residents, who never use it. They also continue to call their home "the moon," much to the surprise of visitors.

EARTH'S MOON IN OUR UNIVERSE

FIRST HUMANS LANDED 1969
238,355 MI (383,595 KM) FROM EARTH
TEMPERATURE RANGE FROM -233° TO 123°C

With minimal weather and geologic movement to remake the lunar surface, our moon presents us with a time capsule of sorts that preserves a physical history written in cratered highlands and immense impact basins for us to examine and explore.

ON JULY 20, 1969, APOLLO 11 ASTRONAUTS NEIL Armstrong and Buzz Aldrin walked on the moon—the first step in our extraterrestrial explorations. Including the Apollo 11 mission, 12 men in total have walked or driven roving vehicles on the moon, all between 1969 and 1972. The prospect of returning to the moon—or traveling even farther, perhaps to Mars—is now back on the drawing boards of NASA and space agencies in other nations: China, India, Russia, Japan, and the European Space Agency (ESA).

A MOON IS BORN

Meanwhile, astronomers on Earth continue to investigate the origins of our moon. The leading theory of the moon's birth suggests that some 4.6 billion years ago, when Earth and all the other planets in the solar system were still forming, a wayward Mars-size planet barreled into ancient Earth. The titanic power of the collision shattered and melted the colliding body and probably much of the outer layer of our planet, too. While the cores of the two worlds merged, the violent impact sent a splash of molten debris into space. It went into orbit, coalesced, and cooled over time to become the moon we see today.

A MOON OF MANY SHADES

The moon orbits around us, changing its relationship between the Earth and the sun. This 29.5-day cycle gives us an ever shifting lunar view. And then there's a lunar eclipse, which occurs when the moon, Earth, and sun come into alignment. During an eclipse, our planet's shadow is projected onto the full moon, slowly darkening its surface until it seems to vanish before our eyes.

The moon's appearance during total lunar eclipse can be a very deep red, rust-colored, or even fiery orange. This variable coloring results from the angle of the moon's path through the umbra—Earth's conical dark shadow, which blocks the sun's light. The atmospheric

ARE WE THERE YET?

MINING THE MOON

***Star Trek*'s Orpheus Mining Colony's insignia states "sine qua non," meaning "without which, nothing." In our universe, NASA is aiming to identify and extract moon resources that could become similarly essential. Lunar Flashlight, set to launch in 2017, and the Resource Prospector, set for 2018, will map and prospect for water ice that could be split into hydrogen and oxygen for fuel.**

Getting There

ENT / "Demons"

combination of water and solid particles, like dust and volcanic ash, filters the sunlight before it's refracted into the umbra, creating a range of colors from light orange to bloodred, depending on the amount of particulate matter in Earth's atmosphere. 

A dramatic illustration of a planetary collision imagines the moon's birth. It's believed that when a Mars-size protoplanet crashed into Earth billions of years ago, rocky debris spun into space and formed our satellite.

FEDERATION SHIPS

Since its founding in 2161, the United Federation of Planets has built and maintained a number of different vessels: Starships meant for traversing the galaxy on military and diplomatic missions, shuttles for visiting the surfaces of planets and transporting supplies, and large cruisers carrying civilians make up the expansive fleet. Ships both past and present have their own unique specifications and abilities, from faster-than-light travel to powerful weapons.

RUNABOUT **Named after Earth's rivers—Danube, Ganges, Yangtze Kiang, and Rio Grande, for example—Deep Space 9's runabouts are smaller than starships but more comfortable than shuttles.**

DELTA FLYER **Designed to handle the Delta Quadrant's harsh environments—like ion storms, an unstable quantum slipstream, and a dark matter asteroid—the *Delta Flyer* is a mesh of Starfleet and Borg technology designed by the crew of the *U.S.S. Voyager.***

U.S.S. DEFIANT NX-74205 **The first of the *Defiant*-class ships, this vessel was assigned to Deep Space 9 to defend against threats to the space station and its inhabitants.**

U.S.S. ENTERPRISE NCC-1701 **A *Constitution*-class starship, the *U.S.S. Enterprise* has served as the Federation's flagship vessel, leading science and discovery missions over 40 years under five captains, most notably Captain James T. Kirk.**

GALILEO NCC-1701/7 **The *Galileo* is a shuttlecraft servicing the *U.S.S. Enterprise* under Captain Kirk's command. After crash-landing on a hostile planet, the crew is forced to abandon the shuttle and return to the *Enterprise* (TOS / "The Galileo Seven").**

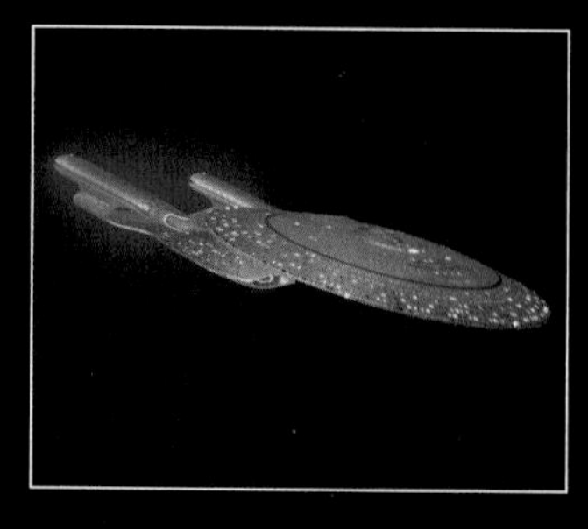

U.S.S. ENTERPRISE NCC-1701-D **Captained by Jean-Luc Picard, this is the sixth Federation starship named *Enterprise*. In addition to standard shuttles, the *Enterprise* has a large, detachable shuttle, the captain's yacht, called *Cousteau*.**

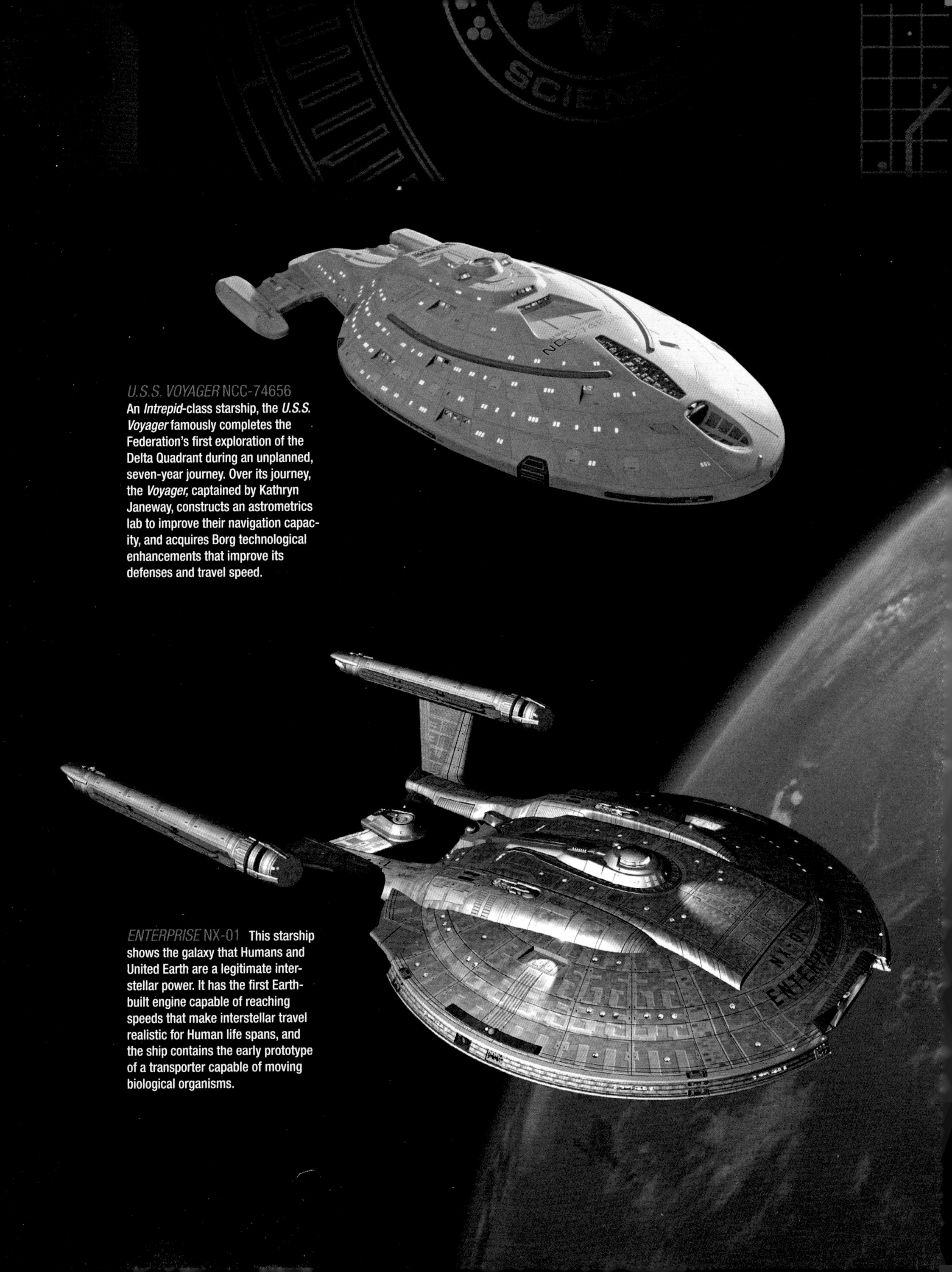

U.S.S. VOYAGER NCC-74656
An *Intrepid*-class starship, the *U.S.S. Voyager* famously completes the Federation's first exploration of the Delta Quadrant during an unplanned, seven-year journey. Over its journey, the *Voyager,* captained by Kathryn Janeway, constructs an astrometrics lab to improve their navigation capacity, and acquires Borg technological enhancements that improve its defenses and travel speed.

ENTERPRISE NX-01 **This starship shows the galaxy that Humans and United Earth are a legitimate interstellar power. It has the first Earth-built engine capable of reaching speeds that make interstellar travel realistic for Human life spans, and the ship contains the early prototype of a transporter capable of moving biological organisms.**

LUNA, THE MOON IN THE NIGHT SKY

Look for Earth's moon, and use unaided eyes or binoculars and telescopes to tour major points of interest on its surface.

STARGAZING TIPS

BEST VIEWING SPOTS: Visible worldwide

BEST TIME TO SEE IT: The best times to observe details on the lunar surface are when features like crater walls and mountains cast shadows, putting them in stark relief. During the full moon, sunlight directly strikes the lunar surface so no shadows are produced. During the moon's waning or waxing periods of its monthly cycle, only a portion of its disk is visible (crescent, quarter moon, or gibbous), and shadows are sharpest around the terminator dark/light line—or "sunrise/sunset" border—offering great views of the craggy landscape.

BASIC TIPS: The moon is the brightest celestial object in the night sky because of its proximity to Earth and because its surface is highly reflective. Looking at it through a telescope can actually tire your eyes quickly. Some amateur astronomers cut down on the moon's glare by using a polarizing or neutral density filter that threads into the bottom of the telescope eyepiece, or by covering a portion of the telescope tube's front opening with a piece of cardboard.

HOW TO FIND IT

1. Look for the moon a couple of days after the new moon phase as it hangs low in the western sky after sunset. Watch it until about two days after first quarter phase. For early risers, the morning skies are great for observing the moon because the air will be less turbulent and views will be clearer starting two days before the last quarter moon phase.

2. Observe the moon with your naked eye, and identify major visible features like the dark impact basins filled with basalt known as "maria" (plural of "mare"). Best known of these is the Mare Tranquillitatis (Sea of Tranquility), the region where Apollo 11 touched down and the first human set foot on another world. Look just above it for the Sea of Serenity and below it for the Sea of Fertility.

3. Use binoculars or a telescope's higher magnification to bring countless impact craters into focus. At the time of the month when the moon is not full, look around the terminator line. One of the most easily recognizable craters is 57-mile (92-km)-wide Copernicus in the mare called Oceanus Procellarum (Ocean of Storms) on the western half of the lunar disk. The crater's bright rays and central mountains are spectacular.

MAP KEY AND SCALE

Lambert Azimuthal Equal-Area Projection

SCALE 1:33,091,000

1 CENTIMETER = 331 KILOMETERS; 1 INCH = 522 MILES

STATUTE MILES 0 250 500

KILOMETERS 0 250 500

* Spacecraft landing or impact site

GRAIL A (Ebb) GRAIL B (Flow) (U.S.) Crashed Dec. 17, 2012
Mare Humboldtanum
J. Herschel
W. Bond
Hiten (Japan) Crashed April 10, 1993
MARE FRIGORIS
Endymion
Sinus Roris
Plato
Aristoteles
Atlas
Sinus Iridum
Chang'e 3 (China) Landed Dec. 14, 2013
Messala
Gauss
OCEANUS PROCELLARUM
Luna 17 (U.S.S.R.) Landed Nov. 17, 1970
Montes Caucasus
Lacus Somniorum
Geminus
MARE IMBRIUM
Posidonius
Russell
Archimedes
MARE
Luna 21 (U.S.S.R.) Landed Jan. 15, 1973
Cleomedes
Apollo 15 (U.S.) Landed July 30, 1971
Luna 15 (U.S.S.R.) Crashed July 21, 1969
Luna 13 (U.S.S.R.) Landed Dec. 24, 1966
Aristarchus
Luna 2 (U.S.S.R.) Crashed Sept. 14, 1959
SERENITATIS
Apollo 17 (U.S.) Landed Dec. 11, 1972
MARE CRISIUM
Montes Apenninus
Menelaus
Palus Somni
Luna 8 (U.S.S.R.) Crashed Dec. 6, 1965
Luna 7 (U.S.S.R.) Crashed Oct. 7, 1965
Luna 5 (U.S.S.R.) Crashed May 10, 1965
Manilius
MARE VAPORUM
Luna 24 (U.S.S.R.) Landed Aug. 18, 1976
Luna 23 (U.S.S.R.) Landed Nov. 6, 1974
Copernicus
MARE TRANQUILLITATIS
Luna 9 (U.S.S.R.) Crashed Feb. 3, 1966
Kepler
MARE INSULARUM
Surveyor 4 (U.S.), Surveyor 6 (U.S.) Crashed, Landed July 17, 1967; Nov. 10, 1967
Ranger 6 (U.S.) Crashed Feb. 2, 1964
Luna 18 & Luna 20 (U.S.S.R.) Landed Sept. 11, 1971; Feb. 21, 1972
Apollo 14 (U.S.) Landed Feb. 5, 1971
Ranger 8 (U.S.) Crashed Feb. 20, 1965
MARE SMYTHII
Surveyor 5 (U.S.) Landed Sept. 11, 1967
Armstrong
MARE FECUNDITATIS
Surveyor 1 (U.S.) Landed June 2, 1966
Apollo 11 (U.S.) Tranquillity Base Landed July 20, 1969
Luna 16 (U.S.S.R.) Landed Sept. 20, 1970
Surveyor 2 (U.S.) Crashed Sept. 22, 1966
Grimaldi
Surveyor 3 & Apollo 12 (U.S.) Landed Apr. 20, 1967; Nov. 19, 1969
MARE COGNITUM
Apollo 16 (U.S.) Landed April 21, 1972
Langrenus
Ranger 9 (U.S.) Crashed Mar. 24, 1965
MARE NECTARIS
Ansgarius
Gassendi
Ranger 7 (U.S.) Crashed July 31, 1964
Vendelinus
MARE HUMORUM
MARE NUBIUM
Byrgius
Petavius
Hecataeus
Palus Epidemiarum
Humboldt
SMART-1 (ESA) Crashed Sept. 3, 2006
Surveyor 7 (U.S.) Landed Jan. 10, 1968
Furnerius
Tycho
Abel
Schickard
Longomontanus
Maginus
Janssen
Phocylides
Clavius
Lyot
Lunar Prospector (U.S.) Crashed July 31, 1999
Selene Kaguya (Japan) Crashed June 10, 2009
Bailly
Chandrayaan-1 Moon Impact Probe (India) Crashed Nov. 14, 2008
LCROSS Centaur Impactor (U.S.) Crashed Oct. 9, 2009

"This is the 32nd planet
I've set foot on!"
—MALCOLM REED

MARS IN *STAR TREK*

ENT / "TERRA PRIME"
SEASON 4 / 2005

IN THIS EPISODE: When xenophobic terrorist John Paxton and his Terra Prime followers threaten to destroy Starfleet headquarters on Earth, the *U.S.S. Enterprise* NX-01 is ordered to fly to Mars and shut down the enemy command center. The *Enterprise* team plans to launch a shuttlecraft with an armed crew led by Captain Archer that can infiltrate Terra Prime's headquarters within the Orpheus Mining Colony. Using an advancing comet as cover, the shuttlecraft reaches Mars without being detected. An intense shoot-out follows, and the *Enterprise* team is able to realign the enemy array beam so that it just misses Earth.

AS THE ENTERPRISE *APPROACHES MARS, THE* rusty red orb looks uncannily Earthlike. With ice-covered polar regions, rocky highlands, and ruddy-hued deserts, the red planet's irregular terrain looks not unlike certain regions of Humanity's home planet. A *Star Trek* shuttlecraft slicing through the Martian atmosphere would pass the Carl Sagan Memorial Station that marks the resting place of NASA's rover Sojourner, a cosmic heritage site that nods to the real-life Pathfinder mission of the 1990s.

Orbiting between Earth and the asteroid belt, *Star Trek*'s Mars is ground zero for Human colonization of deep space beyond Earth and Luna. In the early 2030s, expedition teams tasked with exploring surface topography and drilling for samples touch the first Human boots to Mars's red soil (VGR / "One Small Step"). Early explorations to Mars find the fossils of insects, proving that life can exist there (TNG / "The Last Outpost").

Despite the harsh climate, the first official settlements are established in 2103. Colonists reside in domed cities until the planet is terraformed, making Mars the first planet to be ecologically modified for Human habitation. Verteron arrays redirect comets and asteroids toward the red planet, aiming them at polar caps where the impact will release carbon dioxide. As CO_2 is released into the atmosphere, the planet's temperature and water volume increase; by 2155, Humans can roam freely in the lowlands without heavy environmental suits (ENT / "Demons").

BUILDING A FLEET

The* U.S.S. Enterprise *NCC-1701-D, the* U.S.S. Voyager, *and the* U.S.S. Defiant *were all built in the Utopia Planitia Fleet Yards in orbit around Mars.

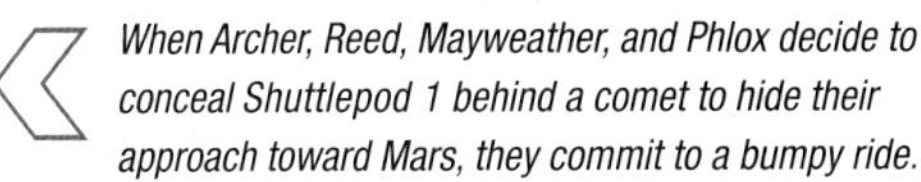

When Archer, Reed, Mayweather, and Phlox decide to conceal Shuttlepod 1 behind a comet to hide their approach toward Mars, they commit to a bumpy ride.

STORM WINDS UP TO 60 MPH (97 KM/H)
1 MARS YEAR = 687 EARTH DAYS

Just as *Star Trek*'s Mars beckons Humans from afar, our real-life Mars is seen as an enchantingly exotic, prize piece of potential real estate. This mysterious red-orange orb seizes our imaginations and embodies our hope for life on other planets.

PART OF MARS'S ALLURE IS THAT WE RECOGNIZE its features: Its polar caps, mountains, canyons, and deserts aren't so different from Earth's. Mars even experiences four seasons, though each one lasts almost twice as long as it does on Earth because of Mars's longer orbital period, or year of 687 days, which means it takes longer to circle the sun.

PROBING MARS

Since the early 1960s, real-life space exploration has launched a robotic invasion of Mars. A flotilla of more than 20 unmanned probes have been successfully launched either to orbit or to land on the red planet. As many and more have failed, whether burned in Mars's atmosphere or unable to maintain contact once in orbit. Of the successes, which include operational crafts from Europe's ESA and India's Indian Space Research Organization (ISRO) in addition to NASA, several continue to bring us valuable information from the planet's surface or from above. Yet arriving is only half the battle: The fine dust that covers Mars is often whipped into dust storms that have been clocked at up to 60 miles an hour (97 km/h). It can take months for the dust to settle, making mapping efforts a particular challenge.

Explorations of Mars have focused especially on evidence of water, a vital ingredient in the recipe for life. Its surface is etched with canals and branch systems, floodplains and basins, likely made by rivers and streams some billions of years ago. Recent evidence points to the probability that Mars and Earth looked very much the same four billion years ago.

In 2012, NASA's Mars Curiosity rover landed in a giant crater believed to be a former lake and found conglomerate rocks made of small, rounded pebbles that mirror the form of Earth's waterworn rocks, suggesting that water once flowed there. In 2015, the Mars Reconnaissance Orbiter confirmed changes over time in a portion of the planet's landscape that prove the current existence of flowing water.

ARE WE THERE YET?

POPULATING MARS

By the 24th century, *Star Trek*'s Mars is a thriving Human hub, with citizens like Dr. Leah Brahms (at right). In 2015, we found evidence of life-giving water on Mars, but how close are we to settling there? Applications are already open for eager travelers who want to apply to take part in the Mars One program, which hopes to have humans settled on the red planet by 2027.

Getting There

TNG / "Galaxy's Child"

Scientists are devising ways to make small, chilly Mars habitable. Among the issues yet to be resolved is how to shield long-term Martian residents from the effects of solar radiation.

MARTIAN ART

The** Enterprise **NCC-1701-D's guest quarters features a painting of both Mars and one of its moons, Deimos—the Latin word for "terror" (TNG / "Heart of Glory").

DREAMING OF AN OASIS

The long-term vision of Mars as a human-friendly oasis is gaining support, but there are many hurdles to habitation. Mars can reach as high as 70°F (21°C) at the midlatitudes, and can plummet to minus 225°F (-143°C) at the poles. Combined with its atmosphere of 95 percent carbon dioxide, the conditions would cause unprotected astronauts to suffocate and their blood to boil due to low atmospheric pressure.

One proposed method of creating a habitable environment involved harvesting gases trapped in the Martian soil to increase the atmosphere's carbon dioxide levels. Recent observations from the MAVEN mission showed Mars has lost much of its atmosphere to solar winds and that any gases further artificially released would likely be lost in the same way. Scientists must continue to dream up other means of making Mars livable.

CURIOSITY "SELFIE"

This self-portrait of NASA's Curiosity Mars rover shows the vehicle at the "Mojave" site, where its drill collected the mission's second taste of Mount Sharp, visible on the horizon. The scene combines dozens of images taken during January 2015 by the Mars Hand Lens Imager (MAHLI) camera at the end of the rover's robotic arm. The pale "Pahrump Hills" outcrop surrounds the rover, and darker ground holds ripples of windblown sand and dust.

MARS IN THE NIGHT SKY

Look for Mars, the orangish red-hued starlike planet in the sky, traveling along the ecliptic.

STARGAZING TIPS

BEST VIEWING SPOTS: Visible worldwide

BEST TIME TO SEE IT: Choose nights with low humidity and haze, when the stars twinkle less. Observe the best Mars views late at night, when the ruddy planet rides high in the sky, above Earth's atmosphere's worst blurring effects. Count on seeing the planet at its brightest and biggest in our sky when it comes close to us every 26 months, known as opposition (May 2016, July 2018, October 2020, December 2022, January 2025, and February 2027).

BASIC TIPS: Of all the planets visible in the sky, Mars and Mercury are the only ones whose surfaces are not shrouded in clouds, allowing us to see some details. But to observe the red planet's features will require backyard telescopes with 4- to 6-inch mirrors using high-power eyepieces working at least 170 times or more magnification. The most conspicuous Martian features are its bright, white polar caps and dusky shadings on its surface. Larger telescopes can at times reveal cloud bands and moving dust storms.

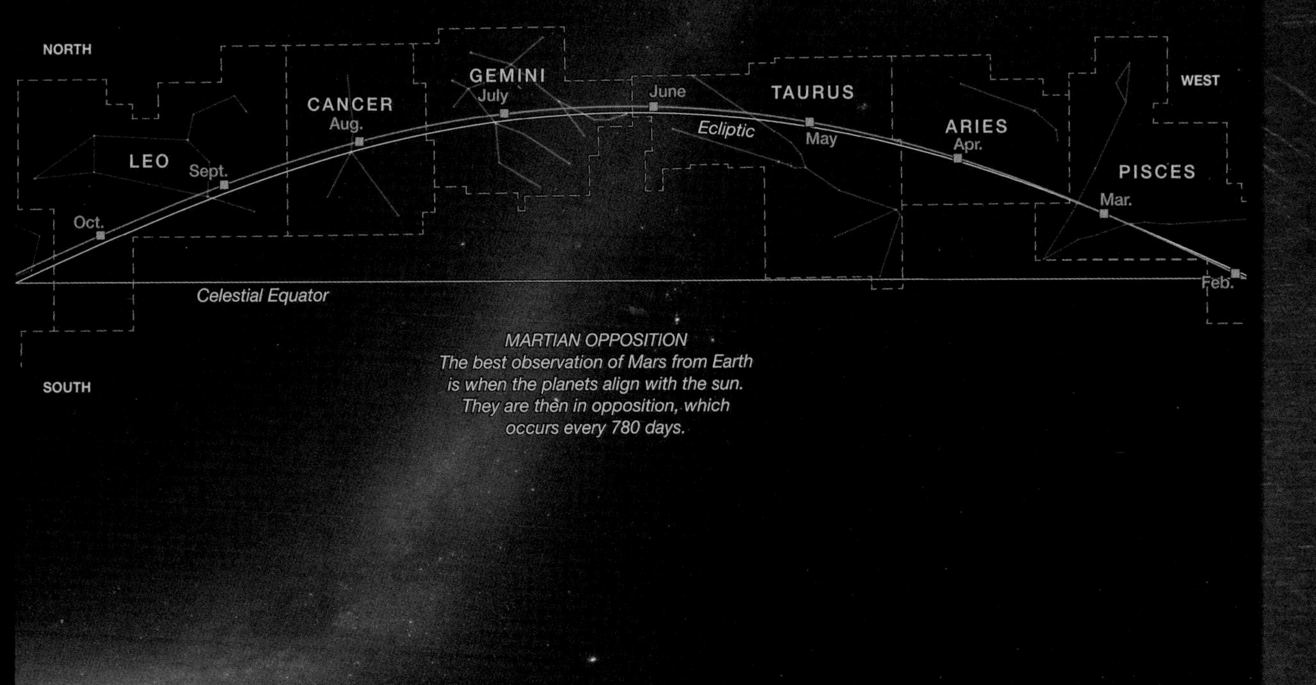

MARTIAN OPPOSITION
The best observation of Mars from Earth is when the planets align with the sun. They are then in opposition, which occurs every 780 days.

HOW TO FIND IT

1. The sky chart above shows the band of the sky that follows the ecliptic line and the constellations strung along it. It also shows the apparent motion of Mars as it moves across the sky over the weeks and months of two Earth years, which is the time it takes the red planet to complete one orbit around the sun.

2. Mars's path is shown in relation to different constellations and is symbolized by a colored line, with each color representing a particular year. A corresponding colored marker shows the position of the planet at the start of each year. Choose the current year and look for the approximate location of the current month along the colored line representing the planet's path in the sky to identify the constellation it lies within.

3. Refer to the seasonal sky charts on pp. 224-227 to find the position of the identified constellation in your sky where Mars will be at that time. If local sky conditions are favorable, the red planet should be bright enough to spot with the naked eye. That distinctive ruddy color you see is caused by sunlight hitting the iron oxide–rich sands and dust of Mars. For more precise sky maps of Mars's positions within specific constellations, go online to TheNightSkyGuy.com.

"I recommend a tour of Jupiter's third moon. I hear the lava flows are lovely this time of year."

—DR. LEWIS ZIMMERMAN

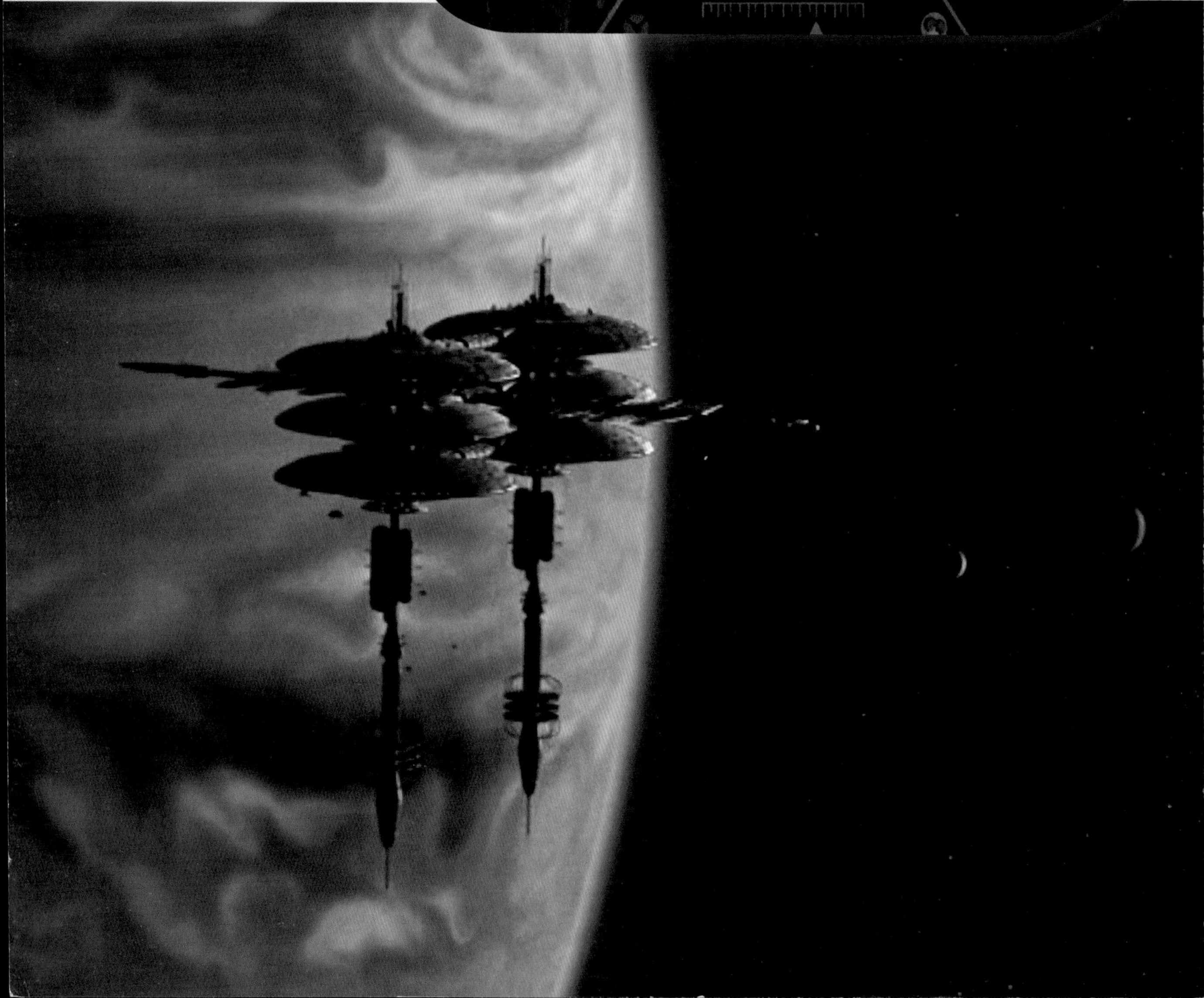

IN THIS EPISODE: Jupiter Station is a maintenance and repair center for Starfleet vessels and home to the cranky, eccentric genius Dr. Lewis Zimmerman, who engineered the *U.S.S. Voyager*'s Emergency Medical Hologram. When *Voyager* receives word from Starfleet headquarters that Zimmerman is terminally ill, the Doctor plots to save him with what he believes may be a cure: an experimental treatment using Borg regeneration techniques. The team compresses a hologram transmission, enabling them to send the Doctor to Jupiter Station some 30,000 light-years away so he can see the ailing Zimmerman—his creator—face to face.

WHEN JUPITER STATION COMES INTO VIEW, THE massive orange planet suspended behind it makes a powerful backdrop. Its psychedelically colored atmosphere features bands of clouds that weave their way around swirling gases. As the Doctor streams to Jupiter Station, the planet's rotation brings the Great Red Spot—a giant, hurricane-like storm—into view. Two of its satellites, Ganymede and Io, are visible off in the distance.

The fifth planet from Sol and the largest in the Sol system, Jupiter is a gas giant situated near the asteroid belt. In the 1970s, Pioneer 10 first surveyed the planet's magnetosphere to see how it would affect objects in its vicinity. By 2151, Jupiter's popular repair and supply station is orbiting the planet, serving starships from near and far (ENT / "Silent Enemy," "Fortunate Son").

Jupiter is a planetary fixture of the *Star Trek* universe. In the opening sequence of *Star Trek: The Next Generation*, the *Enterprise* passes it en route to or from colonies closer to the sun. Jupiter makes another cameo in 2270 when the *Enterprise* speeds toward V'Ger, a massive alien life-form emitting enormous amounts of energy that poses a threat to Earth *(Star Trek: The Motion Picture)*. By the 24th century, Jupiter can be reached thanks to the Jovian Run, a shuttle connecting it with Saturn. Edward Jellico and Geordi La Forge pilot shuttles on it early in their Starfleet careers (TNG / "Chain of Command, Part II"). Tourists purportedly take vacations to Jupiter to enjoy lava flows on one of its moons.

Jupiter Station serves starships in need of repair or enhancements, as well as officers seeking additional training. The view isn't bad, either.

HIDDEN RESOURCES

In* Star Trek Into Darkness, *Scotty finds a top-secret construction hangar—a Section 31 facility—orbiting Jupiter's volcanically active moon Io.

JUPITER IN OUR UNIVERSE

67 CONFIRMED MOONS
CLOUDS 30 MI (48 KM) THICK
1 JUPITER DAY = 10 EARTH HOURS

Dubbed the king of planets for its staggering size, Jupiter measures more than 11 times the diameter of Earth and twice the mass of all the other worlds in the solar system combined. This huge, marbled beauty and its multitudes of moons is just as breathtaking in real life as it appears in *Star Trek.*

THE SHEER MAGNITUDE WITH WHICH JUPITER IS depicted in *Star Trek* can only help us begin to grasp the enormity of the planet in real life. Jupiter could swallow up more than 1,300 of our relatively tiny Earths. Its name is appropriately adopted from the Roman god of the sky. It's located approximately 483 million miles (just over 777 million km) from the sun, beyond Mars and the asteroid belt. So remote is this spectacular planet that it takes an average of about 40 minutes for sunlight reflected off its cloud tops to reach Earth.

UNDER PRESSURE

With an atmosphere 600 miles (almost 1,000 km) deep, composed mostly of hydrogen and helium blanketing its surface, Jupiter is the quintessential gas giant. It spins on its axis so quickly that its clouds are stretched and pulled like taffy into long bands that appear as stripes across its surface. These cloud belts—the two most distinctive circling above and below the planet's equator—are easily visible even through a small telescope. Adding to its mesmerizing appearance is the presence of heavier atmospheric elements like sulfur and compounds like methane and ammonia, which combine to create its vibrant white bands, called zones, and reddish belts.

If you were to descend through its atmosphere, you'd find that the towering clouds get ever denser. Eventually the increasingly immense pressure squeezes them into liquid hydrogen that acts as a metal, measuring about 25,000 miles (over 40,000 km) thick. At the center, the pressure is so high that each square inch of your body would experience more than 650 million pounds of weight, equal to 130,000 cars.

Most of our knowledge of Jupiter's interior atmosphere was gathered during the Galileo mission that orbited the planet in 1994, parachuting a probe that descended more than 93 miles (150 km) into its depths.

ARE WE THERE YET?

SCANNING FOR MALADIES

The medical tricorder is one of Doctor McCoy's handiest tools, capable of scanning and diagnosing a patient's condition in seconds. A *Star Trek*–inspired gadget called the Scanadu Scout has turned McCoy's tool into a reality. Anyone can place this tiny device against the forehead and check blood pressure, heart rate, temperature, and oxygen levels: medical diagnostics in the palm of your hand.

Mission Accomplished

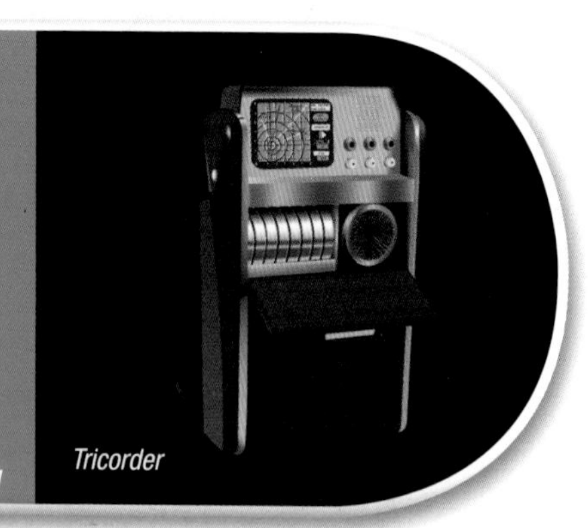

Tricorder

RETINUE OF MOONS

Jupiter's godly status extends to its miniature system of worlds, with 67 moons and counting, as of 2015. Its four largest moons—called the "Galilean moons" after Galileo Galilei, who first spotted them in 1610—are visible through binoculars. The moons themselves are named for Jupiter's mythological lovers Io, Europa, Ganymede, and Callisto, all comparably sized to Earth's moon, but each is wonderfully unique.

Io, the innermost moon, is the most geologically active body in the entire solar system. Ice-covered Europa has water and hints of organic chemicals that suggest it could harbor life. The largest Jovian moon, Ganymede, has a crater-pocked landscape that may be hiding a subsurface ocean like its neighbor Europa. The outermost Galilean moon, Callisto, has been sculpted by collisions. Valhalla, the largest of its "bruises," appears as a bull's-eye formation with concentric ring patterns that span 1,900 miles (over 3,000 km) across.

Jupiter's four largest moons (clockwise from top left): Io, with its sulfurous volcanoes; icy Europa, with its subsurface ocean of water; Callisto, with a diameter nearly identical to Mercury's; and Ganymede, the largest moon in the solar system

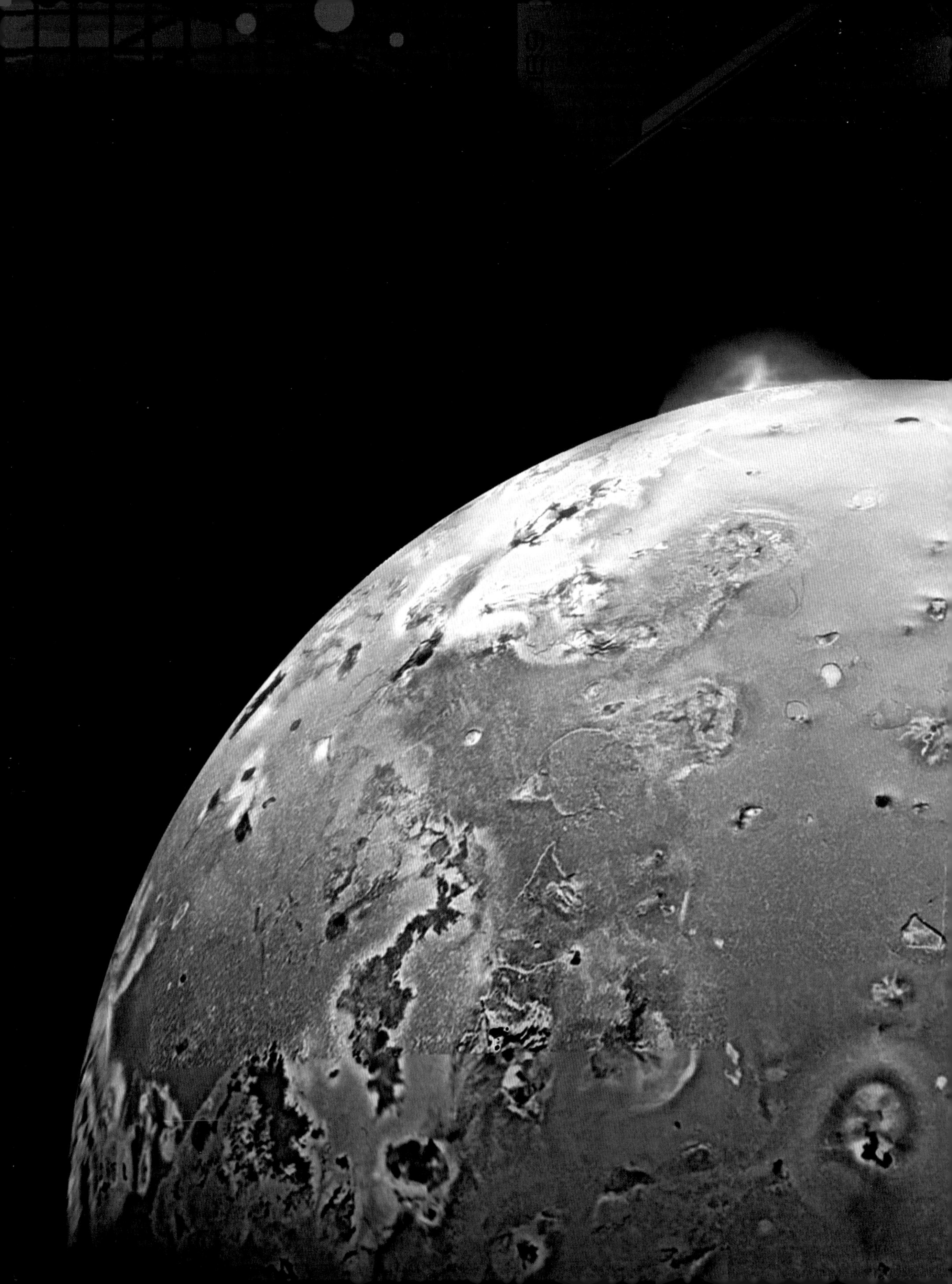

IO'S VOLCANIC ERUPTION

Jupiter's innermost Galilean moon, Io, is the solar system's most geologically active. Here, a bluish plume marks a volcanic eruption that sends particles shooting 60 miles (100 km) above its surface. This composite uses images taken with green, violet, and near-infrared filters of the Solid State Imaging system on NASA's Galileo spacecraft.

JUPITER IN THE NIGHT SKY

Look for Jupiter, the starlike planet with a creamy-white hue, traveling along the ecliptic in the sky.

STARGAZING TIPS

BEST VIEWING SPOTS: Visible worldwide

BEST TIME TO SEE IT: The second brightest planet after Venus, Jupiter appears in our skies like a very brilliant star. That's because it is a huge world—able to fit 11 Earths across its diameter—and reflects a lot of sunlight off its cloud tops. The clearest views of Jupiter are when it reaches its highest point in our skies, and the planet's light has to penetrate the thinnest amount of Earth's thick atmosphere.

BASIC TIPS: Although Jupiter looks like a bright star to the naked eye, steadily held binoculars will reveal it as a bright disk, with its four largest starlike moons (Io, Europa, Ganymede, and Callisto) lined up on either side of the planet, at varying distances from each other and Jupiter itself. Because the innermost moon, Io, takes less than two days to orbit Jupiter, you can actually witness its relative position change in just an hour or so, while its neighboring moons move from night to night.

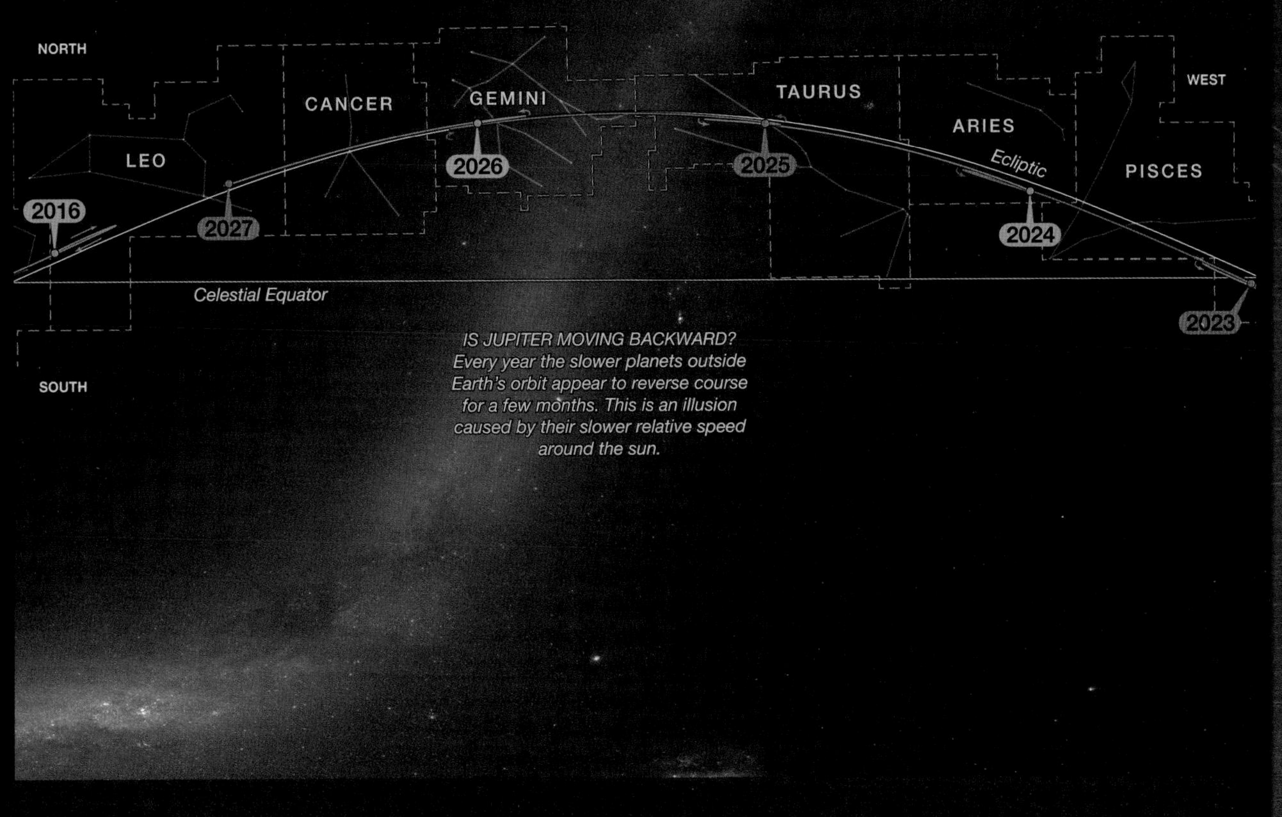

HOW TO FIND IT

1. The sky chart above shows the band of the sky that follows the ecliptic line and the constellations along it. It also shows the apparent motion of Jupiter as it moves across the sky over a 12-year period, the time it takes to complete one orbit around the sun.

2. Jupiter's path is shown in relation to different constellations and is symbolized by various colored lines, with each color representing a particular year. A corresponding colored marker shows the location of the planet at the start of each year. Choose the current year and look for the approximate location of the current month along the colored line representing the planet's path in the sky, and identify the constellation it lies within.

3. Refer to the seasonal sky charts on pp. 224-227 to find the position of the particular constellation in your sky where Jupiter is located at the time of year. If local sky conditions are favorable, the largest planet in the solar system should be bright enough to spot with just the naked eye. For more precise sky maps of Jupiter's positions within specific constellations, go online to TheNightSkyGuy.com.

"Your father was captain of a starship for 12 minutes. He saved 800 lives . . . I dare you to do better."

—CHRISTOPHER PIKE

STAR TREK / 2009

IN THIS MOVIE: Time-traveling Romulan villain Nero attacks planet Vulcan, prompting the Federation to send a rescue flotilla (including the *U.S.S. Enterprise*) under Captain Pike's command. With Captain Kirk and Spock at odds about how to catch Nero—the man responsible for Kirk's father's death—young Chekov hatches a plan to trail him toward Saturn and drop out of warp within the thick atmosphere of Titan, one of its moons, which will hide their ship. The element of surprise allows Spock and Kirk to beam aboard Nero's ship and thwart his attempt to destroy Earth by creating a black hole at Earth's core.

KNOWN FOR ITS MAJESTIC RINGS, SATURN IS A beautiful and iconic part of the *Star Trek* universe. In distant frames it appears a warm ochre yellow with faint cloud bands that alternate with dark, thin concentric gaps.

In the *Star Trek* universe, planetary travelers know that Saturn's largest moon, Titan, has a dense atmosphere that makes it ideal for a ship's concealment. The inner moons remain unidentified, but they surely mirror the planet's real-life moons Rhea, Dione, and Tethys. Pilots flying the Jovian Run shuttle route between Jupiter and Saturn sometimes pull a maneuver near Titan called "Titan's Turn": an atmosphere-grazing feat of daring, dangerous enough that it's been banned (TNG / "Chain of Command, Part II").

Saturn is the sixth and second largest planet from Sol, and it plays muse to many *Star Trek* explorers. As a child, Rain Robinson could see Saturn's rings through her brother's telescope. Thinking they glittered like pirates' treasure, she longed to reach up and touch them, a desire that inspired her to become an astronomer (VGR / "Future's End, Part II").

The* Enterprise *uses the atmosphere of Saturn's moon Titan to hide from the* Narada. *Titan's atmosphere is toxic to Humans, but it is dense enough to provide good cover.

RECITING SATURN

Saturn features in a poem Charles Evans forces Spock to recite: "Saturn rings around my head, down a road that's Martian red" (TOS / "Charlie X").

The first Human spacecraft to reach Saturn, Pioneer 11, dates back to 1979—a *Star Trek* nod to real-world history. Pioneer discovers the planet's magnetosphere, which, like Jupiter, affects objects in its orbit and makes it difficult to scan the planet's surface (TNG / "Loud as a Whisper"). In the 22nd century, another probe spots signs of mining activity on Saturn's second and fourth moons, alluding to an alien history.

SATURN IN OUR UNIVERSE

1 SATURN DAY = 10.7 EARTH HOURS
885,904,700 MI (1,426,725,414 KM) FROM THE SUN / 763.6 TIMES EARTH IN VOLUME

The sixth planet from our sun takes a Federation starship hours to reach, but in our universe it takes a probe the better part of a decade. Despite the distance, scientists behind Saturn imaging missions have given us priceless glimpses of its beauty.

THANKS TO SPACE TELESCOPES AND ROBOTIC probes, Saturn is one of the most recognizable planets in our universe. Its beautiful rings were first observed by Galileo in 1610 and have been the objects of wonder and exploration ever since. As solidly as they have settled into the human imagination, we now know that the rings are made of orbiting particles, primarily water ice. Although they are visible with even the most basic telescope, their thickness measures somewhere between 30 and 300 feet (9 to 90 m). Like Jupiter, Saturn is a gas giant and is less dense than water, which means it could float in a swimming pool if one existed that was large enough to hold it. Saturn and its atmospheric gases rotate so rapidly—a day lasting only 10.7 Earth hours—that its equator bulges out, making it the most flattened planet in the solar system.

SYMPHONY OF RINGS

Theories abound on the rings' origin. One explanation for the icy rings is that their particles are the shattered remains of ancient, small moons that ventured too close to—and were torn apart by—Saturn's gravity. Geysers on one of its moons, Enceladus, feed Saturn's outermost ring ice particles. Scientists suggest we are lucky to live in a time when the rings are visible because they are also temporary, likely to get sucked into the planet or disperse outward and disappear within a few hundred million years.

The latest high-powered imaging has helped us see some of Saturn's phenomena in jaw-dropping detail. We now know that its rings are made of thousands of bright, narrow bands like concentric grooves on a vinyl record, interspersed wit h dark gaps. One of the largest rings, for example, stretches as much as 300,000 miles (almost 483,000 km) out from the planet itself.

MINIATURE SOLAR SYSTEM

Saturn's rings maintain their shape because of the planet's shepherding moons. Its system of more than 62 icy moons of varying sizes and compositions—the

ARE WE THERE YET?

BUILDING AN *ENTERPRISE*

A real-life *Enterprise* may seem like pure science fiction, but an enthusiastic engineer called BTE-Dan thinks we could build one in the next 20 or 30 years. Although some of the technology we'd need for things like gravity wheels and fusion engines are within our grasp, one big complication is power. Without warp drive or something like it, our spaceship won't be traveling far.

Light-Years Away

U.S.S. Enterprise *NCC-1701*

A false-color image of Saturn's rings is used to represent information about ring particle sizes in different regions based on the measurements from the Cassini spacecraft: Purple shows where particles are larger than two inches, green and blue where they are smaller than two inches.

most complex collection in the solar system—pushes and tugs at the rings' particles, helping them maintain their position around the planet's equator. The Cassini spacecraft has been touring Saturn and its domain since 2004, orbiting and re-orbiting the planet to take stock of these diverse moons. Titan—the moon that has a starring role in *Star Trek* (2009)—is the largest of them all, shrouded in a thick orange smog created by sunlight interacting with the moon's nitrogen-rich atmosphere. Cassini's observations show that Titan may be similar to ancient Earth in that they share a substantial atmosphere as well as surface features such as lakes, rivers, dunes, rain, mountains, and possibly volcanoes—all of interest to astronomers.

GADGETS

What would a vision of the future be without an imagination of the technological progress that comes along with it? The universe of *Star Trek* is full of gadgets and games—some outlandish and others not so different from what we use today.

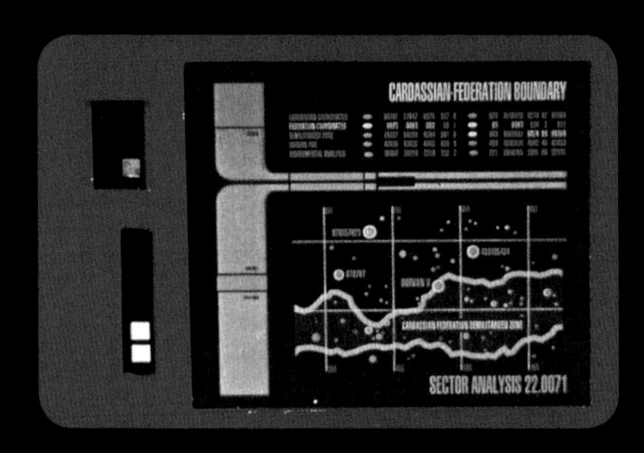

PADD **Short for Personal Access Display Device, the PADD is a handheld computer with a large touch screen. These devices become more powerful and efficient over time, with the PADDs on Captain Picard's *Enterprise* cased in boronite whisker epoxy and capable of surviving a 10-foot (35-m) drop without damage.**

GEORDI'S ENGINEERING KIT
As chief engineer of the *Enterprise*-D, Geordi La Forge is a talented programmer and problem solver. Using his engineering kit, La Forge has repaired countless ships and thought his way out of dangerous situations.

DESKTOP VIEWER **Used across species and planets, the desktop viewer is one of the most common pieces of equipment seen in *Star Trek.* These devices are used for everything from communication to data storage—like the modern computers they closely resemble.**

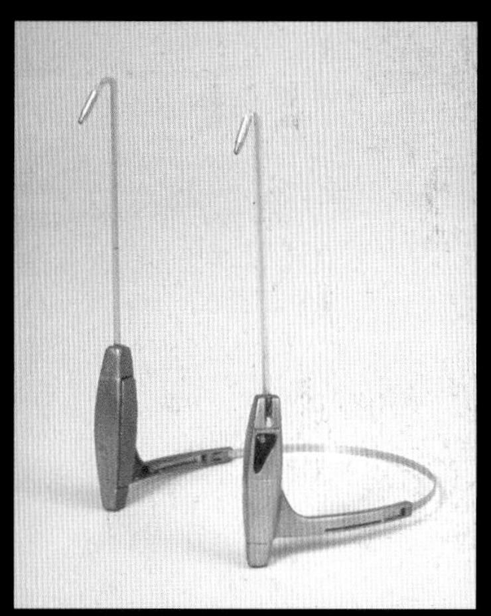

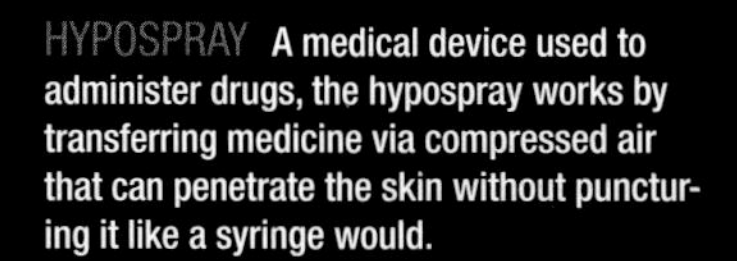

HYPOSPRAY **A medical device used to administer drugs, the hypospray works by transferring medicine via compressed air that can penetrate the skin without puncturing it like a syringe would.**

KTARIAN GAME **Members of Starfleet aren't invulnerable to the addictive properties of mobile games. In fact, the Ktarian game is so impossible to put down that it almost takes out the *Enterprise*-D's crew. Only Data proves immune to the game's neurological scheme.**

SATURN IN THE NIGHT SKY

Look for Saturn, the golden yellow–hued starlike planet in the sky, traveling along the ecliptic.

STARGAZING TIPS

BEST VIEWING SPOTS: Visible worldwide

BEST TIME TO SEE IT: To the naked eye Saturn shines with a steady light and does not twinkle like stars because its beam of light hitting our atmosphere is thick enough to penetrate intact, preventing it from scattering. Saturn takes 29 years to make one trip around the sun, so it travels slowly in Earth's skies, creeping across the background of stars and their constellations. This means that at times in its orbit Saturn is tilted toward Earth, providing great views of its awe-inspiring rings, while other times the planet's rings appear edge-on and nearly disappear from our view.

BASIC TIPS: The sixth planet from our sun looks like a bright star to unaided eyes, but to see its famous rings requires a small backyard telescope working with at least 30 times more magnification than the human eye. The second largest moon in the entire solar system, Titan, can be glimpsed, and a scope with at least a 6-inch mirror will show four of Saturn's other major moons.

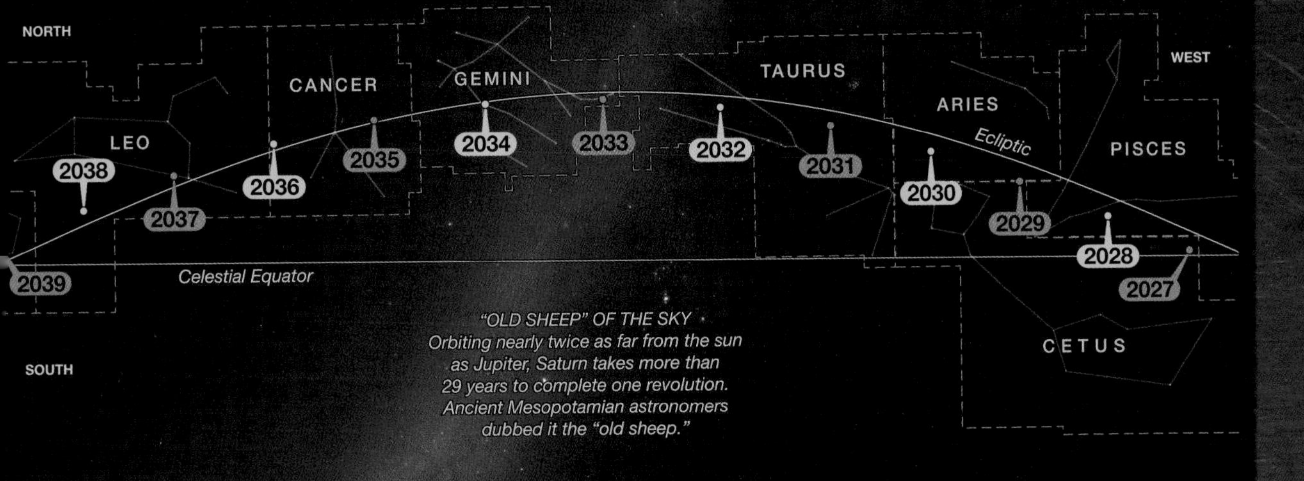

"OLD SHEEP" OF THE SKY
Orbiting nearly twice as far from the sun as Jupiter, Saturn takes more than 29 years to complete one revolution. Ancient Mesopotamian astronomers dubbed it the "old sheep."

HOW TO FIND IT

1. The sky chart above shows the band of the sky that follows the ecliptic line and the constellations along it. It also shows the apparent motion of Saturn as it moves across the sky over a 29-year period, the time it takes to complete one orbit around the sun.

2. Saturn's path is shown in relation to different constellations and is symbolized by various colored lines, with each color representing a particular year. A corresponding colored marker shows the location of the planet at the start of each year. Choose the current year and look for the approximate location of the current month along the colored line representing the planet's path in the sky, and identify the constellation it lies within.

3. Refer to the seasonal sky charts on pp. 224-227 to find the position of the particular constellation in your sky in which Saturn is located for your time of year. If local sky conditions are favorable, the ringed world should be bright enough to spot with the naked eye. For more precise sky maps of Saturn's positions within specific constellations, go online to TheNightSkyGuy.com.

"For those of you who aren't near a window, you might want to find one. There's something pretty amazing off starboard."

—JONATHAN ARCHER

COMETS IN *STAR TREK*

ENT / "BREAKING THE ICE"
SEASON 1 / 2001

IN THIS EPISODE: The discovery of a large comet prompts the *Enterprise* crew to investigate, a Vulcan ship trailing not far behind. Learning that the massive comet—the biggest ever observed by Humans or Vulcans—has rare and potentially useful eisillium beneath its surface, Captain Archer orders a team to take the shuttle pod in to collect samples. The team approaches the comet's head through a thick, hazy atmosphere and lands on a craggy surface lined with treacherous fissures. When drilling begins, an explosion tilts the comet's axis, pushing the shuttlepod toward the star and forcing a rescue mission.

VULCANS MAY SOUND SCORNFUL WHEN THEY CALL comets "nothing more than dirty snowballs," but they are tapping into an evolving scientific understanding of these astral bodies that we share in real life. No matter how we characterize them, mercenaries and explorers in the *Star Trek* universe find many ways to make good use of the comets in our Sol system.

Comets are chaotic and unpredictable objects, which is why the *Enterprise* is instructed to land near the comet's pole, where it is dark and the surface is likely more stable.

In the 22nd century, comets are the primary instrument used to terraform Mars into a habitable planet. In 2155, as many as 14 comets are redirected by verteron array to impact the planet's polar regions within a 30-month period. The goal of these impacts is to release carbon dioxide, intended to warm the atmosphere (ENT / "Terra Prime") to make Mars more hospitable to Humans.

Sometimes a comet's rich contents are put to even more direct use—as in 2377 when Harry Kim, Seven of Nine, and the Doctor collect cometary biomatter to create new medicines (VGR / "Body and Soul"). Comets don't always serve useful purposes, however. In 2370, the *Enterprise* finds an 87-million-year-old D'Arsay cultural archive inside a rogue comet that tries to take over the ship (TNG / "Masks"). Whether useful or not, comets are chaotic and unpredictable objects, which is why in "Breaking the Ice," the *Enterprise* is instructed to land near a comet's dark end, where the surface is likely to be more stable.

The Enterprise *trails soon-to-be-named Archer's comet, hoping to find a way to land on its jagged surface and mine its deposits of coveted eisillium.*

COMETS MATTER

In 2266, the U.S.S. Enterprise *uses a Romulan ship's disturbance of a comet's tail to pinpoint its position and engage it in battle (TOS / "Balance of Terror").*

COMETS IN OUR UNIVERSE

4.6 BILLION YEARS OLD / 3,000+ CURRENTLY KNOWN / THE MOST FAMOUS, HALLEY, VISIBLE TO EARTH EVERY 75 YEARS

A collection of frozen leftovers from the planets' formation, comets carry relics of the solar system like stories from our past. Modern spacecraft let us peer into the hearts of comets, revealing cores as fascinating as *Star Trek* envisions.

WONDERS OF THE NIGHT SKY, COMETS ARE BONA fide members of our solar system that orbit the sun in predictable patterns and are periodically visible to us on Earth. They may look like stars, but they are conglomerates of planetary debris that coalesced into billions of solid formations made up of dust and ice. They were probably born near where Jupiter is now, and then cast out into the frozen depths of the outer solar system—beyond the orbit of Neptune and Pluto. Like the Vulcans in *Star Trek,* astronomers call them "dirty snowballs."

STRANGE ANATOMY

Comets orbit the sun with a more elongated orbit than planets, traveling far from the sun and then boomeranging back in close again. Whenever a comet's heart (or nucleus) nears the sun, its ice starts to melt and evaporate, and dust particles shed off. These particles—some no bigger than a grain of sand—travel at speeds of up to 124,000 miles an hour (200,000 km/h), and when it happens that our planet plows through the orbiting debris field, we on Earth experience it as a meteor shower.

In 2005, NASA's Deep Impact successfully blasted a crater in the nucleus of comet Tempel 1, revealing a spongy interior with lots of holes. These holes could mean the comet was formed from many so-called dirty snowballs, with ice as the glue holding it together. A dark, carbon-based material makes the comet's surface appear black. It's coated with a fine layer of dust whose tiny particles stay separate because the comet is too small to have enough gravity to consolidate them.

Just like in *Star Trek,* landing on a comet is a tricky business. When the European Space Agency sent their Philae lander to comet 67P/Churyumov-Gerasimenko in 2014, they had to drop it from 14 miles (22.5 km) away onto a comet traveling 40 times faster than a speeding bullet and covered in cracks, fissures, and potentially a thick layer of dust.

ARE WE THERE YET?

TRACTOR BEAMS

Ubiquitous in *Star Trek,* tractor beams let starships move objects remotely in space using elementary particles called gravitons to transmit the force of gravity. The real world is catching up: Scientists can now use sound waves to levitate objects, and NASA has teamed with Arx Pax (of real-life hover technology fame) to build a device that can manipulate satellites without actually touching them.

Getting There

TNG / "The Battle"

Comet Hale-Bopp was one of the brightest comets of the 20th century, visible to Earth during the first half of 1997. The comet has two clear tails: a white dust tail and a blue ion tail that both point away from the sun.

A TRAIL OF CRUMBS

With every trip around the sun, comets release more debris. Particles like dust and chunks of rock can form a growing train of castoffs in the comet's wake that persists for thousands of years. Halley's comet, for instance, is estimated to have a life span of 350,000 years, thus making 4,600 orbits around the sun.

During a shower, the meteor streaks appear to come from the same point in the sky. They are thus named for the constellation from which they radiate: For example, the Perseids emanate from the constellation Perseus and the Geminids from the constellation Gemini. At least a dozen showers can be viewed on known dates each year.

COMET OF DOOM

In 2371, the U.S.S. Defiant *encounters a comet some hail as the "Sword of Stars," prophesied to destroy the Celestial Temple (DS9 / "Destiny").*

ROSETTA COMET WATCH

The Rosetta spacecraft is carrying out an unprecedented investigation of the nucleus and environment of comet 67P/Churyumov-Gerasimenko. The orbiter, a large aluminum box with two enormous solar panel "wings" extended from the side, carries a suite of scientific instruments for data collection and a lander, Philae. The image below was captured 17.6 miles (28.3 km) from the comet's center in 2015, when comet rotation and slightly different illumination revealed a wealth of surface details on a previously hidden side.

COMET 67P **Comet 67P/Churyumov-Gerasimenko is named after its discoverers, Klim Churyumov and Svetlana Gerasimenko. The first observation took place in 1969, during a survey of comets by astronomers in Kazakhstan. It measures just under two miles (3.2 km) wide and three miles (4.8 km) long.**

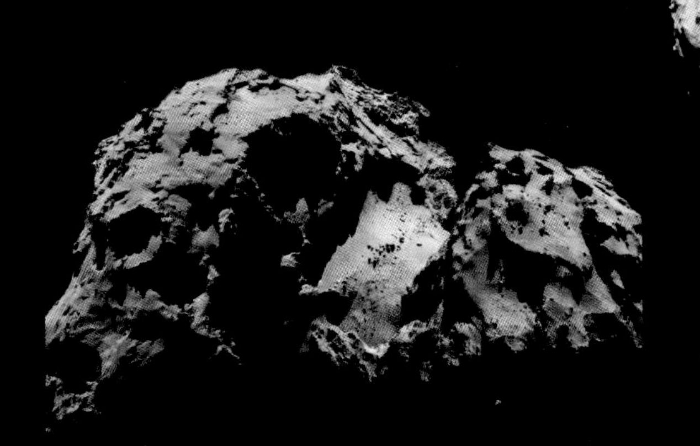

A SAD FAREWELL **The Rosetta mission's lander, Philae, came to rest in a shaded location that prevented its solar-powered secondary batteries from recharging. It collected valuable data while operational, sending intermittent signals until July 2015. In February 2016, the European Space Agency declared that Philae was in "eternal hibernation."**

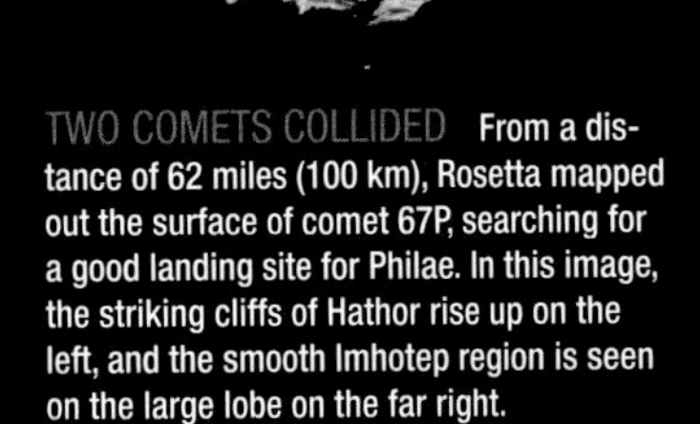

TWO COMETS COLLIDED **From a distance of 62 miles (100 km), Rosetta mapped out the surface of comet 67P, searching for a good landing site for Philae. In this image, the striking cliffs of Hathor rise up on the left, and the smooth Imhotep region is seen on the large lobe on the far right.**

PHILAE'S RESTING PLACE
Philae is about the size and shape of a washing machine. During its descent to comet 67P's surface, it failed to deploy its anchor harpoons and was bounced back into space twice, before eventually coming to rest in a less-than-ideal shaded location.

ACTIVE PITS **OSIRIS, the main imaging system of the Rosetta mission, consists of two cameras: one wide-angle, and one narrow-angle. This image, showing 18 pits in comet 67P, was taken from 177 miles (285 km) away by the narrow-angle camera. Some of those pits are active.**

METEOR SHOWERS IN THE NIGHT SKY

Look for meteor showers and the parent constellations from which they appear to radiate.

STARGAZING TIPS

BEST VIEWING SPOTS: Most are visible worldwide.

BEST TIME TO SEE IT: In the Northern Hemisphere, meteor showers peak between midnight and predawn. Time your observations for a less-bright moon phase. More meteors are visible when the shower's parent constellation rises above the horizon.

BASIC TIPS: The best observation location for meteor showers is the countryside, far from city light pollution. It can take up to a half hour for your eyes to adapt to the dark, and any stray light can ruin night vision in seconds. Watch for meteors at least halfway up the sky. Position yourself facing away from the shower's radiant, and you'll be ready to make a bunch of wishes.

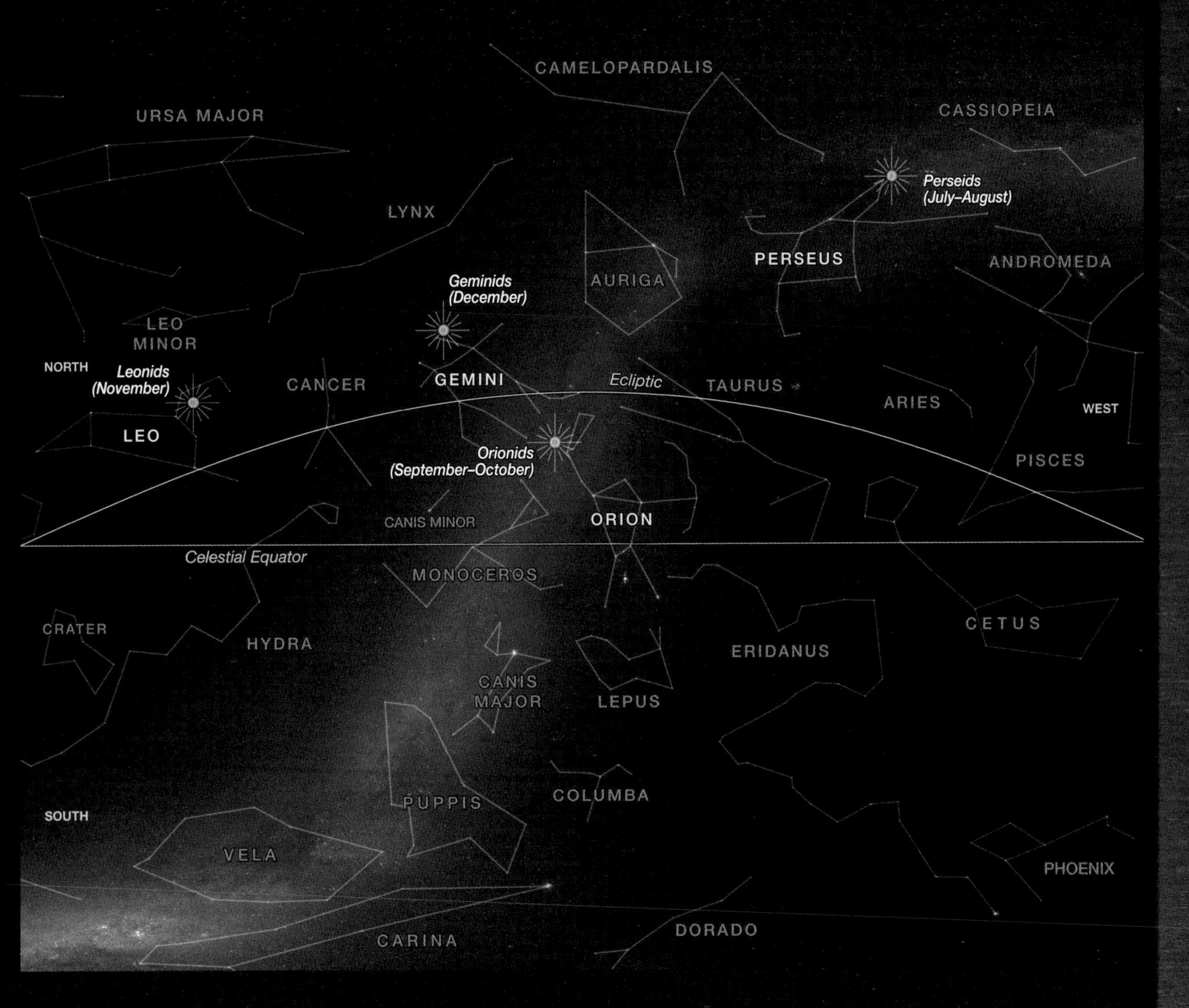

HOW TO FIND IT

1. The all-sky chart highlights the brightest constellations visible across all seasons. It also shows the various radiant points of major meteor showers located within specific constellations.

2. The date and time of annual meteor showers are very predictable. Refer to the list of annual meteor showers and their expected dates of peak activity, which usually span a couple of nights of observing. Each shower appears to always emanate from the same point in the sky, so find the radiant of the shower within its namesake constellation on the corresponding sky chart. Check the seasonal sky charts on pp. 224-227 to find the position of the particular constellation in your sky for the time of year.

3. Don't worry if you miss a shower's peak time; shower activity can last as much as a week before and after the peak night. Meteors zip across large tracts of the overhead sky in a fraction of a second, so the naked eye is the best way to watch these cosmic fireworks shows.

NCC-1701
NCC-1701

STRANGE NEW WORLDS

DEEP SPACE MISSIONS ARE FINDING THOUSANDS OF CURIOUS PLANETS BEYOND OUR SOLAR SYSTEM.

CHAPTER TWO

STAR TREK & US

To tour the *Star Trek* universe, you would have to journey thousands of light-years—a distance beyond your wildest imaginings, teeming with planets spread across four different quadrants. Our solar system is a comparative pinhead, a tiny fraction of the vast expanse to be explored. Of the countless stars in the *Star Trek* sky, a rich diversity of planets is orbiting them. In our reality, such planets beyond our solar system are called exoplanets, but we have yet to see them in the vivid detail with which *Star Trek* worlds are portrayed.

Some *Star Trek* planets are barren and desolate, like Delta Vega and Gemulon V; others—like jungle-dense New Earth and rain-soaked Ferenginar—are exotic worlds where alien species thrive. Others still are home to sophisticated colonies and settlements like the Federation outposts Calder II and Caleb IV. In the *Star Trek* universe, these networks of inhabited worlds evolve into civilizations dominated by alliances and coalitions, as well as conflict and outright war. Could such societies flourish in the far reaches of our universe?

CATEGORIZING THE COSMOS

The landscapes of distant planets in the *Star Trek* universe are not wholly unfamiliar: there's the rocky, crater-rich terrain of Mercury and our moons; the windswept deserts of Mars and oceans characteristic of Earth's; the bands of colorful clouds that encircle gas giants like Jupiter; and the spectacular concentric rings reminiscent of Saturn's. There are dense jungles, too, although the flora and fauna that thrive within them are often unlike anything on Earth.

Star Trek's United Federation of Planets uses a planetary classification system that's far more refined than anything we've needed so far. It identifies how habitable or safe a world is based on detailed physical properties like size, atmospheric composition, and the presence of life. These planetary classes are designated by letters—more than 20 classes exist in the *Star Trek* universe, but this list represents eight commonly explored types:

D Asteroids and moons (Regula, Paan Mokar)
H Uninhabitable desert worlds (Tau Cygna V)
J Gas giants (Jupiter, Saturn)
K Livable in atmospheric domes (Mars, Theta VIII)
L Habitable, but only plant life (Indri VIII)
M Breathable atmosphere, lots of water (Earth, Vulcan, Bajor)
N Barren, sulfuric, thick atmosphere (Venus)
Y Toxic atmosphere, radiation, and high temperatures (demon worlds)

PREVIOUS PAGES: The original Enterprise *spent decades discovering fascinating worlds beyond our solar system, some reminiscent of Earth—and some just the opposite.*

SUPER-EARTH GLIESE 667 Cc

An artist's rendering shows a sunset as seen from the super-Earth Gliese 667 Cc. The brightest star in the sky is the red dwarf Gliese 667 C, which is part of a triple-star system. The other two more distant stars, Gliese 667 A and B, also appear in the sky to the right.

Starfleet is always interested in exploring class M planets, where conditions are suitable for humanoid life. As astronomers continue to discover potentially Earthlike exoplanets in our galaxy, they have developed a simple classification system to categorize and describe them that includes three subcategories—subterran, terran, and superterran—for types we would think of as class M or habitable planets.

SEARCHING FOR EXOPLANETS

The quest for exoplanets continues apace in our universe. For much of the 20th century, the idea of detecting relatively tiny planets orbiting stars light-years away seemed impossible. The planets could be seen only by the starlight they reflected from their host stars, and that light would appear billions of times fainter than the light emanating from the star itself.

IN SEARCH OF ALIEN WORLDS

NASA launched the Kepler mission in 2009. Its orbiting vehicle houses a photometer, or light-sensitive telescope, trained to search for planets outside our solar system—such as those shown in this artist's vision, some of which are Earthlike and possibly habitable. Thousands of candidates have been identified, and the search continues.

A breakthrough occurred in 1992 when scientists detected two planetary masses orbiting a pulsar, or a magnetized neutron star. In 1995, a third exoplanet, orbiting a main sequence star, was confirmed. Since then, the rate of discovery has skyrocketed; by early 2016, the tally of confirmed exoplanets had reached nearly 2,000.

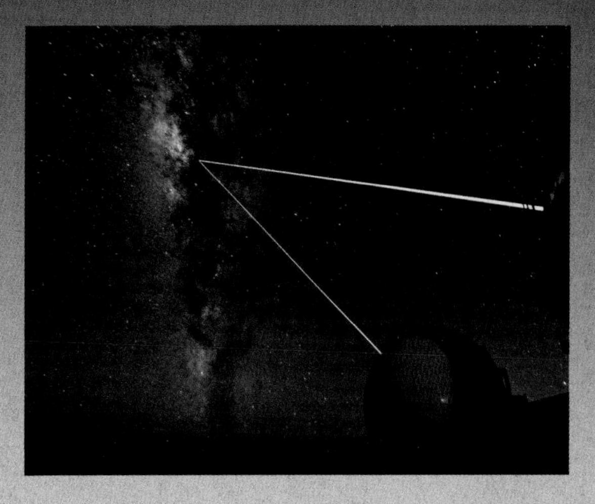

The twin Keck telescopes shoot guiding lasers toward stars in the heart of the Milky Way on a beautifully clear night on Mauna Kea's summit.

The technology for identifying these planets is racing to catch up with *Star Trek*. The "transit method," by which the orbiting planet blocks a tiny bit of starlight as it passes in front of—or transits—its host star, is challenging because the star and exoplanet must align directly in the observer's line of vision. It's easier to see big planets than smaller planets, because they block more of the star's light. But with technology progressing year by year, new tools and search methods are gathering paradigm-changing knowledge about what's out there. NASA's Kepler mission, launched in 2009, is finding more exoplanets than we ever dreamed existed.

A WEALTH OF WORLDS

The *Star Trek* universe is so profuse with life that a protocol for meeting new races or governments—known as "first contact"—is firmly in place. The Federation only initiates contact if a civilization is deemed sufficiently advanced and capable of interstellar travel. If they aren't sufficiently developed, the Prime Directive strictly prohibits Starfleet from interfering with them—a mandate that gets Kirk into hot water when he violates it at the planet Nibiru in order to save Spock *(Star Trek Into Darkness)*. In the *Star Trek* universe, Humanity's definitive first contact was their encounter with Vulcans on April 5, 2063, when the *T'Plana-Hath* touched down in Bozeman, Montana, introducing Humans to worlds beyond our own *(Star Trek: First Contact)*.

First contact might seem far-fetched in our universe, but teams and observatories around the globe are hard at work searching the galaxy for traces of life—or sometimes just the ingredients required for life as we know it. Exoplanet hunters seek habitable worlds with life-sustaining temperatures, estimated largely by the brightness of the host star and the planet's distance from it. Unlike boiling-hot planets like Mercury and freezing-cold planets like Neptune, exoplanets with temperatures like Earth's reside in a "Goldilocks zone" where we hope to find evidence of extraterrestrial life, or at least the possibility of it.

It's estimated that as many as a billion of the planets in the Milky Way could be rocky, Earth-size planets with conditions suitable for liquid water on the surface—class M planets for the real world—orbiting stars similar to our own.

"It is said thy Vulcan blood is thin. Are thee Vulcan? Or are thee Human?"

—T'PAU

TOS / "AMOK TIME"
SEASON 2 / 1967

IN THIS EPISODE: As the *U.S.S. Enterprise* speeds toward Altair VI, Spock begins behaving strangely. Spock confides that he is under the spell of Pon farr, the Vulcan mating season. McCoy runs a series of medical tests, which reveal that Spock will descend into madness and die within eight days if he doesn't return to his home planet. Spock, McCoy, and Captain Kirk beam down to Vulcan, where Spock seeks to mate with T'Pring, who prefers a Vulcan named Stonn. She claims her right to ask Spock to fight for her, and chooses Kirk as her champion. The crewmates are pitted against each other in a fight to the death.

YOU APPROACH A VAST RED DESERT FRINGED BY craggy mountains. Volcanoes spill lava down their slopes while waves crash onto the shore of a nearby sea. It's bright and very hot. Welcome to the planet Vulcan, an Earthlike planet some 16 light-years from Earth.

Vulcan's landscape is not unlike parts of our planet, but the environment is harsher. The atmosphere is thinner, so you might find it harder to breathe. You will feel heavier, too, with the higher gravity. You'll meet a lot of Vulcans here: the highly logical, pointy-eared aliens who became Humanity's "big brother" when they first introduced them to life beyond Earth. The Vulcans' physiological differences allow Vulcans like T'Pol to save the *Enterprise* crew from phenomena that don't affect them (ENT / "Singularity").

Planets suitable for humanoid life are designated "Minshara" (class M) in the *Star Trek* universe, a label that the Vulcans originated and that the *Enterprise* crew first used to describe the planet Talos IV (TOS / "The Cage"). Class M planets tend to have a nitrogen- and oxygen-rich atmosphere, abundant surface water, and a mineral-rich landscape (VGR / "Once Upon a Time").

LIVE LONG AND PROSPER

Leonard Nimoy invented Spock's iconic Vulcan salute (inspired by Jewish hand gestures) to make his character's people unique.

Class M planets in the *Star Trek* universe range in age from three to ten billion years and in diameter from 6,000 to 10,000 miles (9,600 to 16,000 km). Class M planets host complex ecosystems with plants, animals, and, in some instances, humanoid species (TNG / "Angel One"). In the 23rd century, there are a predicted three million of these planets in the Milky Way.

The Enterprise *heads toward Vulcan, home to an ancient, intelligent humanoid species who are among the first to develop and utilize warp drive technology.*

NEARLY 2,000 PLANETS, 500 MULTIPLANET SYSTEMS DETECTED / CLOSE TO 5,000 CANDIDATES UNDER EXPLORATION

Starting in 2009, NASA's planet-hunting spacecraft Kepler set forth on the search for Earth-size worlds in the Goldilocks zone—that is, planets orbiting at the perfect distance from a star to support liquid water, and thus life.

IN 2015, NASA SCIENTISTS INTRODUCED THE WORLD to Kepler-438b and Kepler-442b. Both at first seemed top candidates as habitable exoplanets: Both orbit their stars in the habitable zone, close enough that water on their surface would be liquid, an essential condition for life to exist. Both appear to be rocky planets, another life-supporting characteristic.

Found 475 light-years away from Earth, Kepler-438b circles a red dwarf—a star type that is smaller and lower in temperature than our sun. If Kepler-438b were at the right distance from its sun, astronomers hypothesized, it could harbor life. But further observations showed that this red dwarf shoots out solar flares so violent and frequent that they likely destroy any atmosphere around the planet—a discovery that bumps it from our list of habitable exoplanets.

The other life-worthy candidate, Kepler-422b, is farther away—1,651 light-years from Earth—but in other ways is more Earthlike. Its sun is an orange dwarf, brighter than a red dwarf but not as hot or bright as our sun. Kepler-422b's orbit time equals about 4 months, or 112 days. Its diameter is about one-third that of Earth's, and it receives about two-thirds the amount of light from the star it orbits. Researchers give it a 97 percent chance of having liquid water.

OUTER LIMITS OF LIFE

Understanding the host star's properties is essential to determining the planetary characteristics and evaluating the habitable zone in that system. For small red dwarf stars like the one that 438b orbits, the habitable zone might hug close—but not too close or the orbiting planets are fried. For gigantic hot stars, the band of congenial temperatures for life is more distant—but not too distant or the planets are in deep freeze.

As the confirmed exoplanet count climbs, scientists can compare and contrast the sizes of the planets, their orbital periods, and the types of stars that they orbit to glean ever more insight into the capacity for life. Once a planet has been detected by the transit

ARE WE THERE YET?

WHOLE NEW WORLDS

We're learning new things about alien worlds thanks to the Kepler spacecraft. In fact, we've found what seems like the most potentially habitable planet thus far: Kepler-452b, which orbits about the same distance from its star as we do from ours. To measure anything on 452b, we'd need to get up as close as *Star Trek*'s explorers do—currently impossible, because it's 1,400 light-years out of reach.

Light-Years Away

Kepler-452b

The Kepler mission's search area is a 3,000 light-years-wide swath located 500 to 3,000 light-years away from Earth, in the direction of the constellations Cygnus and Lyra.

method, its orbital size can be calculated based on how long it takes to orbit once around the star—sometimes up to 1,000 days—and the mass of the star using Kepler's third law of planetary motion. The planet's size is found from the depth of the transit (how much the brightness of the star drops) and the size of the star. The tally of terrestrial planets that are half to twice the size of Earth is on the rise, with other life-worthy notables like Kepler-452b discovered in 2015.

WARP SPEED

It takes four days for a starship traveling at warp speed to get from Earth to Vulcan, 16 light-years away **(Star Trek: The Motion Picture *[1979]*).**

SPACE BASES

Space stations and starbases are Starfleet's permanent support facilities spread across the galaxy, providing essential military and civilian functions such as ship maintenance, commerce, research and development, training, and recreation. These stationary structures often orbit a planet, though some free-floating stations exist as well.

STARBASE 84 **A space dock maintained by the Federation, Starbase 84 is a massive space station with room for a large number of starships. Starbase 84 is responsible for handling major starship refittings. At one point, the crew of 84 installed a new warp core on the *U.S.S. Enterprise* NCC-1701-D.**

DEEP SPACE STATION K-7 K-7 is a civilian outpost and trading center near the Klingon border. In one incident, an interstellar trader introduces a litter of tribbles—adorable, rapidly reproducing life-forms—that quickly overtakes the station.

JUPITER STATION Orbiting Sol system's largest planet, Jupiter Station provides maintenance and supplies to Starfleet vessels and houses the Jupiter Station Holoprogramming Center, where the Emergency Medical Holographic program was developed.

STARBASE 1 A space port in Earth's orbit, Starbase 1 provides docking for starships doing business on the planet.

STARGAZING

AN EARTHLIKE PLANET IN THE NIGHT SKY

Look for Keid, a potential Vulcan-like planet, within the constellation Orion.

STARGAZING TIPS

BEST VIEWING SPOTS: Mid- to lower latitudes of the Northern Hemisphere, including the southern United States and all of the Southern Hemisphere

BEST TIME TO SEE IT: Early evening, January to April

BASIC TIPS: Keid appears to the unaided eye as a faint yellow +4.4 magnitude star that lies 16.4 light-years away, just visible from light-polluted suburban backyards. It sits in the constellation Eridanus, which is referred to as the river in many cultures because of the flowing pattern its stars follow. The second white component of this triple star system, +9.5 magnitude 40 Eridani B, is visible through binoculars. Ultra-faint +11.1 magnitude, red-colored 40 Eridani C requires a small telescope to glimpse.

HOW TO FIND IT

1. Face the southern sky and look for the great Greek mythological hero Orion—recognizable by the line of three bright stars that forms his belt. The two bright stars to the north will be his shoulders and the two to the south are his knees.

2. When Orion appears clearly in the sky, look to the bright stars Saiph and Rigel that mark his knee and foot, respectively.

3. Moving from left to right (west to east), draw an imaginary arc connecting the bright stars (Saiph and Rigel) and continue following this line until you hit the faint naked-eye +4.0 magnitude star Beid in the constellation Eridanus. This line from Orion to Eridanus is about 20 degrees—equal to approximately the width of two fists side by side, held at arm's length. Just underneath 122-light-year-distant Beid, approximately 1 degree south, is slightly fainter Keid, or 40 Eridani, the triple-star system that contains the mythical Vulcan homeworld.

Cursa β
ω
μ
ν
o¹ Beid
o² Keid
Rigel
Rana δ
ε Sadira
η Azha
π
Zaurak γ
ERIDANUS
τ¹
τ² Angetenar
τ³
τ⁴
τ⁵
τ⁶
τ⁷
τ⁸
τ⁹
Theemin υ¹
υ² Beemin
υ³
υ⁴
υ⁵
υ⁶
υ⁷
Acamar θ
ι
κ
χ
Achernar α

"The probe surveyed a class 6 gas giant this morning. I think we should send it back for a closer look."
—KATHRYN JANEWAY

GAS GIANTS IN *STAR TREK*

VGR / "EXTREME RISK"
SEASON 5 / 1998

IN THIS EPISODE: Determined to retrieve a special probe that was sent into the atmosphere of a gas giant to thwart potential hijackers, the crew of the *U.S.S. Voyager* decides to build a rescue vessel designed by Tom Paris. They race to build the *Delta Flyer* before a nearby Malon freighter can finish its own shuttle, block their launch, and steal the probe for itself. To get at the probe, the *Delta Flyer* must navigate a layer of hydrogen and methane roughly 6,000 miles (10,000 km) beneath the gas giant's surface, withstanding potentially crushing pressure.

GAS GIANTS ARE EXOTIC—EVEN DANGEROUS—places of gargantuan size and tumultuous climate. Ranging from roughly 31,000 to 87,000 miles (50,000 to 140,000 km) wide, gas giants in the *Star Trek* universe are composed primarily of hydrogen and helium, just like stars. They also contain the base gases fluorine, methane, and ammonia. As with our universe's Jupiter and Saturn, gas giants in the *Star Trek* universe are in the colder zone of a star's ecosphere where their gaseous layers increase in temperature, density, and pressure toward the center. Although they're made of the same materials as stars, they lack the mass to ignite nuclear fusion.

STARSTRUCK

In the Star Trek *universe, two colliding gas giants create a fusion reaction that gives rise to a new star (TNG / "Ship in a Bottle").*

Beyond the Sol system's gas giants—Jupiter, Saturn, Uranus, and Neptune—farther-flung giants offer Starfleet diverse solar neighborhoods to explore. Andoria, neighboring the Vulcan system and home of the blue-skinned, humanoid Andorians and its subspecies the Aenar, is a class M moon that orbits a gas giant. This icy moon has a breathable oxygen-nitrogen atmosphere and holds underground cities that draw energy from geothermal activity (ENT / "United," "The Aenar," and "These Are the Voyages").

The intense magnetic fields characteristic of gas giants emit strange waveforms that disturb any radio signals within range. Ensign Travis Mayweather calls the sounds "siren calls," remembering how his dad would relay the noise over the *E.C.S. Horizon*'s speakers (ENT / "Sleeping Dogs").

The Voyager *crew uses the brand-new* Delta Flyer, *which combines Starfleet and Borg technology, to rescue a probe from a gas giant's turbulent atmosphere.*

GAS GIANTS IN OUR UNIVERSE

KEPLER LAUNCHED 2009 / OFTEN DETECTED AROUND METAL-RICH STARS

Although there are only four gas giants in our solar system, astronomers are discovering many outside of it. As of 2015, the majority of exoplanet candidates (1,542) were Neptune-size; 808 were Earth-size; 1,233 were super-Earth-size; 260 were Jupiter-size; and 49 were even larger.

WHILE WE ARE DISCOVERING THAT EXOSYSTEMS do not necessarily follow the patterns of our own solar system, we continue to ask the question of whether the exoplanetary systems might be able to support life, even though their array of planets is different from our own. Planetary orbit, distance from the host star, and characteristics of the host star all affect a planet's makeup and behavior. Scientists continue to ask what universal rules govern these variations—or do the rules change outside of our own solar system?

UNDER PRESSURE

Although we haven't developed the technology to safely explore a gas giant's inner layers as they have in *Star Trek*, astronomers have developed ways to test our understanding of these superdistant planets. In 2014, scientists at the National Ignition Facility (NIF) attempted to re-create the massive pressures of a gas giant here on Earth in order to study the effects on certain materials. The NIF used enormous lasers from an ultramodern machine contained in an aluminum chamber so exotic looking that it was featured as the *Enterprise*'s warp core in *Star Trek Into Darkness*. To simulate the dense material at the core of giant exoplanets, the alien-looking 33-foot (10-m)-wide machine put high pressure on a synthetic diamond—made of crystallized carbon, like some planetary cores—to see how much it would compress. The result was an almost fourfold compression, an insight that can help scientists better understand the mathematical relationship between a carbon planet's mass and density.

TAKING STOCK

As we continue to refine our knowledge of exotic exoplanets, we've been able to categorize Kepler planet candidates by size to help us look for behavioral and compositional patterns in other parts of the universe.

ARE WE THERE YET?

ORBITAL SPACE DIVING

Many Starfleet officers know the terrifying joy of diving toward a planet at hair-raising speeds. Our universe's Felix Baumgartner set a world skydiving record in 2012, sailing through the stratosphere at Mach 1.25. Now there's the RL Mark VI space diving suit, whose special design will allow near-future divers to make jumps from near-space, suborbital space, and even low Earth orbit.

VGR / "Extreme Risk"

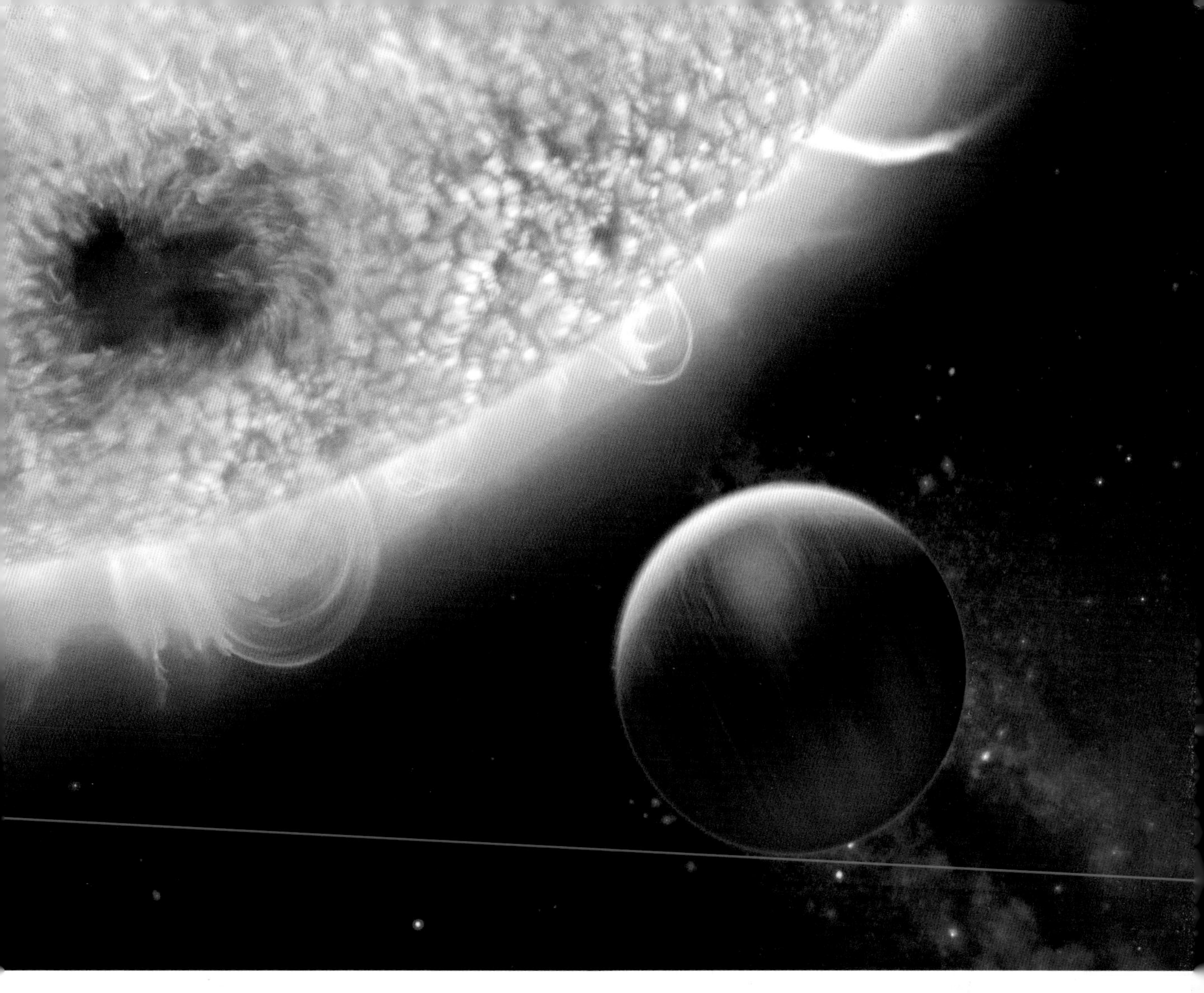

A vivid rendering of extrasolar planet HD 189733 b, now known to have methane and water vapor in its atmosphere. Astronomers detected the molecules by examining light from the host star filtered through the planet's atmosphere.

Among the discoveries from our exoplanet inventory are numerous "hot Jupiters": extra-large gas giants that are extremely close to their parent stars. Based on ongoing studies of our solar system, scientists speculate that as systems mature, planets may experience changes in their orbits. More compact solar systems with planets that experience tidal locking (too-stable axes of spin that keep one side constantly facing its host planet) and have smaller orbits around their parent star would be less likely to be habitable. Most of them have a complex atmosphere, giant storms, and little of the climate variability (that is, night and day or summer and winter) that makes a planet more hospitable to life.

TRIPPY RIDE

Neelix's cousin once enjoyed moving disulfides from a gas giant because of the turbulence, but perhaps also because of the disulfides' hallucinatory effects (VGR / "Friendship One").

A WHOLE NEW WORLD

Inspired by telescope imaging and data, this illustration shows a gas giant planet forming in the ring of dust around the young star HD 100546, whose light is scattered in the thick atmosphere. The system is suspected to contain another large planet orbiting closer to the star.

GAS GIANT EXOPLANET IN THE NIGHT SKY

Look for Pollux, host star of Thestias, a potential Jupiter-like planet, within the constellation Gemini.

STARGAZING TIPS

BEST VIEWING SPOTS: Northern and Southern Hemispheres. From mid-northern latitudes Pollux rides highest in the southern sky in February.

BEST TIME TO SEE IT: Visible evenings, January to May

BASIC TIPS: Pollux, the host star to the giant exoplanet, is a superbright yellow 1st magnitude star that sits 33.8 light-years away from Earth and is easily visible with the naked eye, even from cities. Its home constellation is Gemini, and so Pollux sits next to its mythological twin star, 50.9-light-year-distant Castor. Sky watchers will find that the moon and planets trek through Gemini at times. Uranus and Pluto were both discovered while traveling through the constellation.

HOW TO FIND IT

1. Face the eastern sky and look for a large gathering of bright stars. These represent some of the dominant evening constellations of the season. You'll find an arc of four bright stars, pinned down on the ends by Capella and Procyon: Capella is the brightest member in the constellation Auriga, the Charioteer, and Procyon is the main star in the constellation Canis Minor, the Small Dog.

2. Draw an imaginary arc between Capella and Procyon, and at about the halfway point between the two stars, you will find the Gemini twins, Castor and Pollux.

3. Capella lies about 30 degrees from Castor, equal to the width of three fists side by side, held at arm's length. Procyon is a bit closer—only 20 degrees away from Pollux. The stellar twins themselves are separated by only 5 degrees, equal to about the width of three middle fingers. Pollux shines with a distinct yellow color compared to white Castor.

Capella
θ
Castor α
τ
Pollux β
ι
υ
Jixin κ
G E M I N I
ε Mebsuta
δ Wasat
μ
Nuhatai
η Propus
ζ Mekbuda
ν
λ
γ Alhena
ξ Alzirr
Betelgeuse
Procyon

"We spot any more creatures like that and we'll earn our Exobiology badges."

"Actually, I already have that one."

—ARCHER AND REED

DWARF AND ROGUE PLANETS IN *STAR TREK*

ENT / "ROGUE PLANET"
SEASON 1 / 2002

IN THIS EPISODE: The *U.S.S. Enterprise*'s sensor system detects a lone, drifting planet and decides to investigate. T'Pol leads an away team to survey the rogue planet, where they find a dense junglelike environment that lives in everlasting dark. The Eska surprise them in the forest, explaining that they've been visiting Dakala to hunt wildlife here for generations. While some of the *Enterprise* crewmembers are wary of hunting, they decide to join in. Captain Archer is haunted by a ghostly Wraith who appears in the form of a beautiful woman to beg him to end the hunting, eventually returning to her true reptile-like form.

SOLITARY ORPHAN PLANETS ARE DWARF PLANETS that escaped from their parent star system, gravitationally untethered and orbitless. Their remoteness means they lack a sun's warmth and light, plunging them into eternal nighttime. Nevertheless, Dakala supports a rich ecosystem thanks to steam vents that heat its atmosphere.

Captain Archer compares Dakala's thick vegetation to the rain forests in New Zealand where he earned a merit badge as an Eagle Scout. When the *Enterprise* team meets the Eska and learns about their ritual hunt, their purpose rankles crewmembers who know that hunting has long been out of fashion on Earth. For the Eska, this no-man's-land territory is the perfect place to stalk prey under the cover of darkness, using sensing cloaks to shroud their presence from their quarry.

Dakala's flora and fauna are unusual for a rogue planet, even in the *Star Trek* universe. Geologic activity—like the release of hot interior gases on Dakala—is the only way such isolated planets can host life. Starships rarely encounter these outliers, though in 2371 the *Voyager* approaches one believed to be replete with dilithium—the element used to power warp drive. When Kim, Chakotay, and Neelix beam down to the planet for a geologic scan, they find that Vidiian aliens are already using the planet as a secret organ harvesting facility (VGR / "Phage").

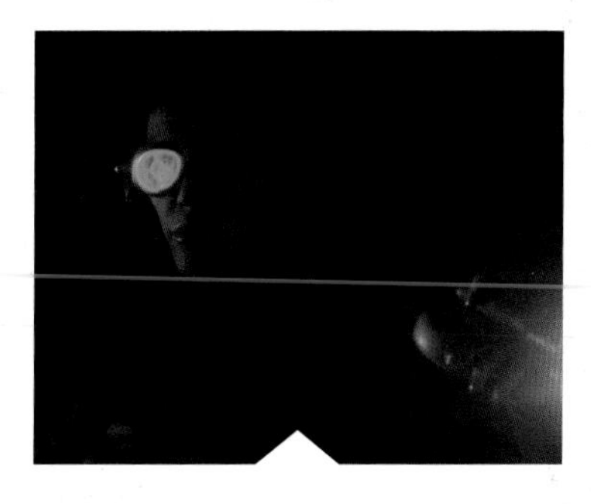

NIGHT SIGHT

This is the only time the crew of the Enterprise *use their green-glowing night vision gear. They're usually dealing with system planets, so they never need them.*

The Enterprise *orbits the ever dark rogue planet Dakala. This planet's lack of a light-emitting host star makes its dense vegetation a puzzle to the crew.*

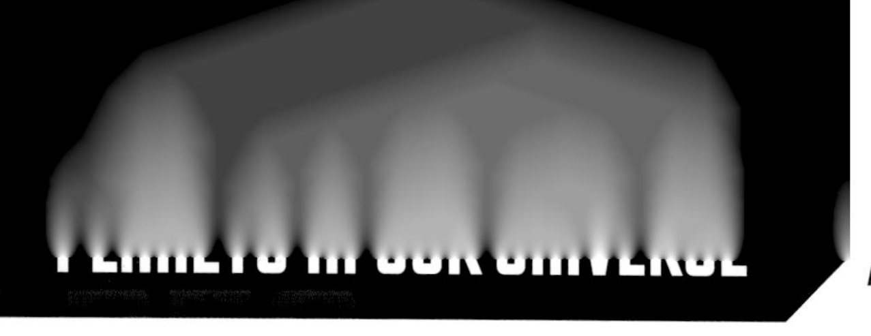

Small dwarf planets can be fascinating science targets, as proven by the recent New Horizons mission to Pluto. Finding new dwarf exoplanets to study is incredibly challenging, though, as their distance from the host star, low mass, and low surface reflectivity can make them difficult to spot.

THE WORLD WAS INTRODUCED TO THE CONCEPT OF dwarf planets in 2006, when petite Pluto was stripped of its status as a major planet and reclassified as a dwarf planet. A need for the designation arose when the International Astronomical Union (IAU) struggled to define the word "planet" as we began discovering objects in the outer solar system—specifically in the Kuiper belt that extends beyond Neptune—that are larger than, or of similar size to, Pluto.

DEFINING DWARFS

By a definition that's still evolving, dwarf planets orbit a star and have enough mass that gravity has made them round in shape. But unlike the major planets, they haven't cleared the orbital path of other bodies in their "neighborhood"—in other words, they're not gravitationally dominant. Pluto, Ceres, and Eris are the first three dwarf planets identified in the solar system, but there could be more than a hundred similar objects lurking in the Kuiper belt. Technology is evolving to give us the tools to spot dwarf planets far beyond Pluto. In 2015, we found dwarf planet candidate V774104, which is small enough to fit inside Pluto twice and lives nine and a half billion miles (around 15 billion km) from the sun. Right now, it's the most distant space object we know of.

LONELY WORLDS

Even more elusive to the eye than dwarf planets are rogues—dwarf planets that were cast away from their parent star during the chaotic birth of their native solar system. For decades these planets were theoretical—seen in the *Star Trek* universe as dark, mysterious worlds that sometimes support strange and surprising life-forms (ENT / "Rogue Planet"). Advanced detection methods could discover starless planets that are at least 300 times the mass of Earth. Scientists are using the best search methods to look

ARE WE THERE YET?

STEALTH MODE

When it comes to evasion, a vital tool in a starship's arsenal is the ability to disguise itself from view and from sensors. It turns out cloaking may not be that hard: Scientists at the University of Rochester have created a 3-D cloaking device using lenses available at any optics shop. Meanwhile, the U.S. military relies on planes that are invisible to radar detection.

Mission Accomplished

The B-2 Spirit

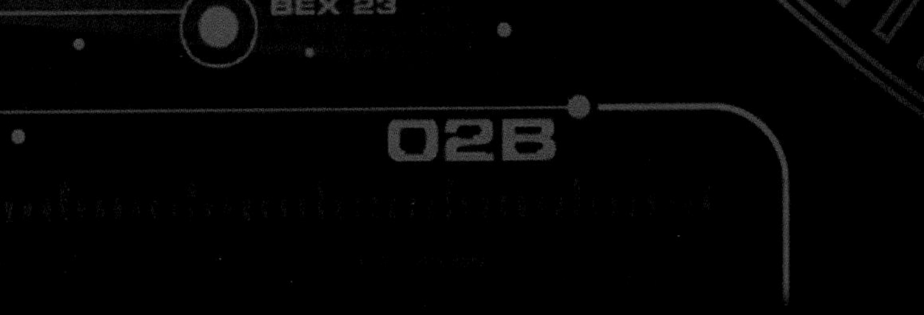

LARGEST KNOWN OBJECTS BEYOND NEPTUNE'S ORBIT

Trans-Neptunian objects are dwarf planets and comets that orbit the sun beyond Neptune's orbit. The largest have enough mass, and therefore gravity, to compress them into a sphere.

for a young rogue's heat in the infrared spectrum or noting its location thanks to "gravitational microlensing," by which the visible light from a distant star is affected by the gravitational field of a closer star or planet.

Not everyone is convinced of the existence of rogues. Some suggest that the objects claimed as rogues are just very remote from their stars or are actually low-mass brown dwarfs that failed to ignite nuclear burning in their core.

DWARF VISITATION

In 2376, the U.S.S. Voyager *visits Theta-class planetoid Norcadia Prime, known for its fine beaches and distinguished museums (VGR / "Tsunkatse").*

ORPHAN PLANET

A 2006–2007 planet survey conducted by New Zealand's Microlensing Observations in Astrophysics suggests that free-floating planets without a host star may be more common than stars in our Milky Way. The lone Jupiter-like planet illustrated here might have been "booted" or ejected from a developing solar system.

EXO-ASTEROID BELT IN THE NIGHT SKY

Look for Vega, host star of an asteroid belt, within the constellation Lyra.

STARGAZING TIPS

BEST VIEWING SPOTS: Across the entire Northern Hemisphere in summer and far northern sky in the Southern Hemisphere in winter months

BEST TIME TO SEE IT: Evening, May to October

BASIC TIPS: Vega, the host star of an asteroid belt, is a blue-white star that shines at magnitude 0, making it the fifth brightest star in the entire sky. Only 25 light-years from Earth, it is one of our sun's closer stellar neighbors, and its brilliance dominates the small constellation Lyra, the Harp. Lyra hosts many other favorite deep-sky stargazing targets like binary and variable stars and one of the most famous star remnants, called the Ring Nebula.

HOW TO FIND IT

1. On late nights in July, look toward the high southern sky for a trio of the brightest stars that form a distinct pattern known as the Summer Triangle. Altair, the lead star in the constellation Aquila, the Eagle, is the closest to the horizon, and despite being the nearest of the three to Earth at only 17 light-years, it is the faintest in the triangle.

2. The next brightest star to the upper left of Altair is Deneb, the most prominent member of Cygnus, the Swan, which is some 3,000 light-years away.

3. Rounding out the trio and riding highest in the sky is the brightest of the three stars, Vega. Its home constellation Lyra is small and compact, consisting of four additional fainter stars in a parallelogram pattern that should be visible from city suburbs under clear skies.

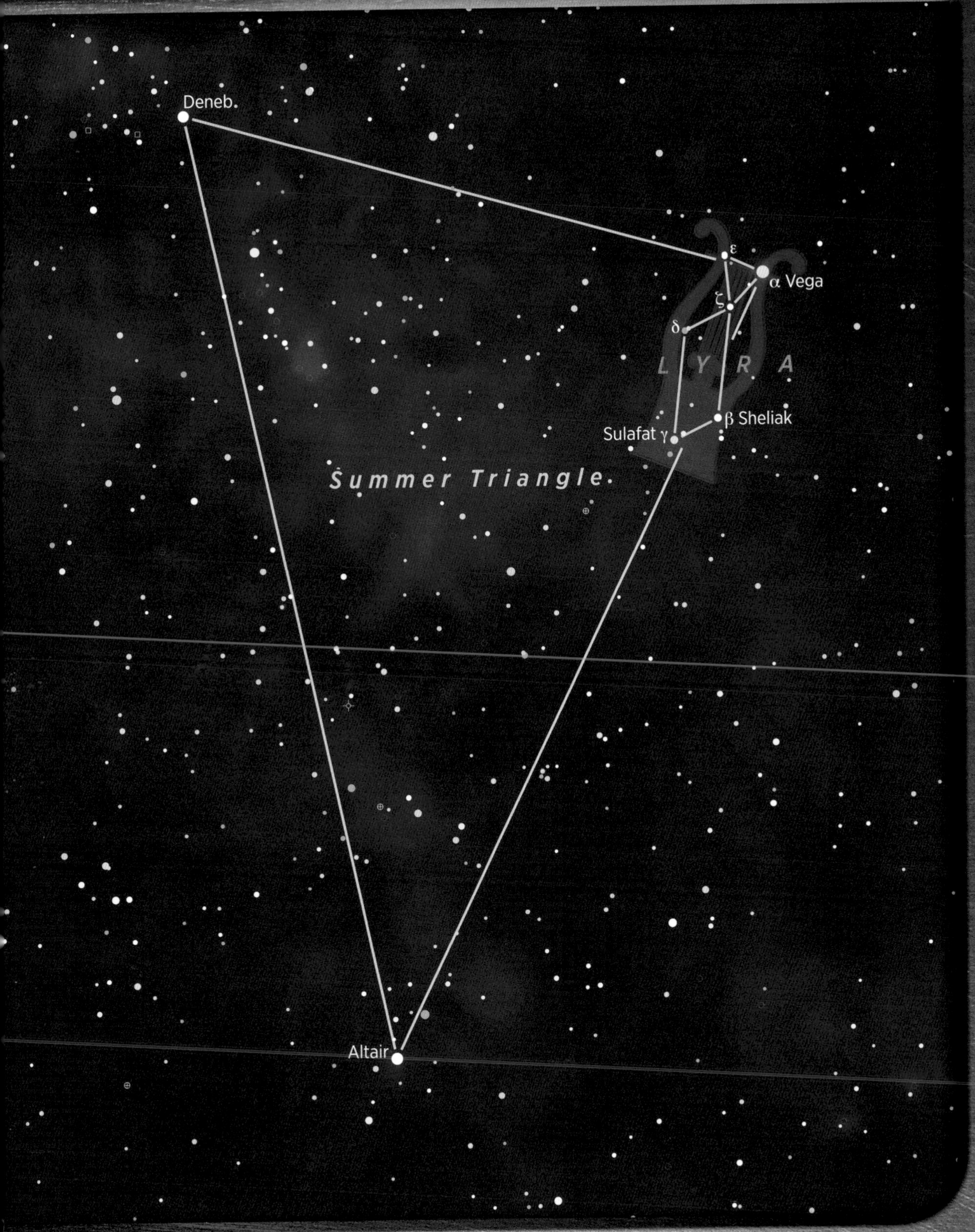

Deneb

ε

α Vega

ζ

δ

L Y R A

β Sheliak

Sulafat γ

Summer Triangle

Altair

“I want these things off my ship! I don’t care if it takes every man we’ve got—I want them off the ship!”
—JAMES T. KIRK
KA

ALIEN LIFE IN *STAR TREK*

TOS / "THE TROUBLE WITH TRIBBLES" / SEASON 2 / 1967

IN THIS EPISODE: The *U.S.S. Enterprise* arrives at Deep Space Station K-7 after being called to protect a shipment bound for Sherman's Planet, whose control the Federation and the Klingons are disputing. Amid tension, a trader arrives bearing tribbles: a fuzzy creature that Uhura delightedly adopts as a pet. The rapidly reproducing fur balls overtake the ship, eating everything they can find and burrowing into the controls. To solve their tribble problem, the *Enterprise* beams the fluffy saboteurs onto a Klingon vessel.

THE SOFT, CUDDLY TRIBBLES AREN'T YOUR STEREO-typical aliens. Their characteristics inspire affection: They coo when touched, a sound that soothes the humanoid nervous system. They increase their numbers exponentially through asexual reproduction, bearing an average litter of 10 tribbles every 12 hours. On their homeworld, predators keep the population in check. But in sheltered environments like the K-7 space station, their unchecked propagation wreaks havoc.

Star Trek's universe is home to a dizzying array of life-forms, some with bodies and some without (called "noncorporeal," or bodyless, species). Noncorporeal species are made of coherent gas or energy and have evolved beyond the need for sustenance, like the Beta Renner cloud that journeys from planet to planet feeding on pain and fear (TNG / "Lonely Among Us").

Most *Star Trek* aliens have bodies made primarily of carbon- or silicon-based cellular material. Then there are the telepathic Medusans, who hide in opaque carrier pods to interact with other species (TOS / "Is There No Truth in Beauty?"), and Changelings, whose natural state is a gelatinous blob that can morph into a physical being of their choosing for 18 hours at a time (DS9 / "The Begotten").

An equally enthralling assortment of nonsentient life-forms—plants, algae, fungi, and microorganisms like viruses and bacteria—populate the *Star Trek* universe. From supersize vegetables to spiny desert creatures, these species turn our definition of "life" on its head.

Deep Space Station K-7 is where Uhura meets shady dealer Cyrano Jones. Within his trove of black market treasures is a purring fluff ball called a tribble.

SAVING KIRK

In "Trials and Tribble-ations," the Deep Space Nine crew goes back in time and saves Captain James Kirk from assassination while he deals with the tribbles (DS9).

ALIEN LIFE IN OUR UNIVERSE

EXTRATERRESTRIAL LIFE ESSENTIALS: TEMPERATURE ALLOWING FOR LIQUID WATER, ENERGY, CARBON

As astronomers continue to find scores of Earthlike planets, astrobiologists speculate as to how extraterrestrial life might look. We are inclined to look for life-forms similar to those we know, but perhaps there are beings out there like nothing we have ever imagined.

ARE WE ALONE? THIS IS PERHAPS THE MOST timeless and universally tantalizing of questions. *Star Trek* tells us we're in the company of a far-flung array of species and worlds. It reminds us of the reality that Earth is not the center of the solar system; that the solar system is not in the center of the Milky Way galaxy; and that the Milky Way galaxy is not in the center of the universe. We could well be living amid wholly different life systems.

LIVING SMORGASBORD

The diversity of life on Earth alone is staggering—from algae to tulips, hummingbirds to snakes—and everywhere we turn are living, breathing organisms, both large and small. *Star Trek*, too, gives us a biological treasury of awe-inspiring variety: Denebian whales, Ligorian mastodons, moon grass, Alvanian bees, and Talaxian tomatoes. The possibilities seem almost endless.

We call the study of life elsewhere in the universe "exobiology," a field that focuses on the chemical variables that create and support life. We know that the most common elements in the universe are hydrogen, helium, oxygen, and carbon in order of abundance. Because helium is inert, the three most abundant, chemically active ingredients in the cosmos are also the top three building blocks of life on Earth. We are going on the assumption (for now) that life on another planet would be made of a similar mix of elements.

EVOLUTIONARY CLUES

But what might that life look like? For now, researchers investigate evolutionary history on Earth for clues. Looking at the molecular record in living organisms and the geologic record, they can tell when and where life first appeared and what the earliest living organisms were like physiologically. Of particular interest are "extremophiles"—organisms that thrive under extreme conditions, or at least conditions unsuitable for familiar life-forms, such as an atmosphere without oxygen, for example, or in

ARE WE THERE YET?

NICE TO MEET YOU

The aliens we meet probably won't be as sophisticated as the ones on *Star Trek*. But let's say they are: How will we communicate with them? *Star Trek* has universal translators (right); we have Microsoft's Skype Translator, which translates languages in real time. It may not help us understand non-Earth languages, but such technology could take us a step in the right direction.

Getting There

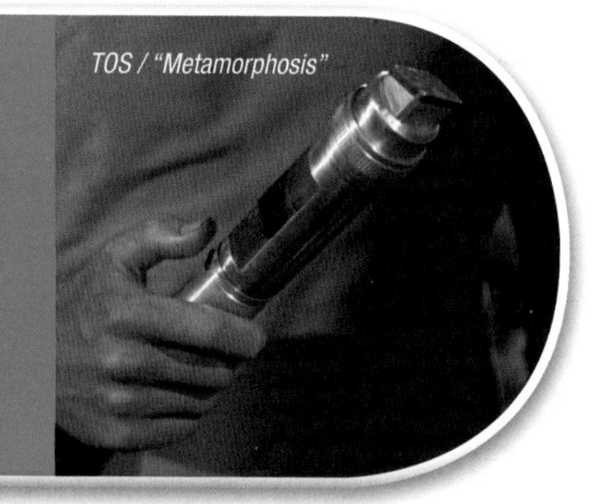

TOS / "Metamorphosis"

A millimeter-long tardigrade—or water bear—shown via a color-enhanced scanning electron micrograph might be Earth's best alien candidate thanks to its extraordinary survivability: The invertebrates can go decades without food or water and endure extreme temperatures and pressures as well as direct exposure to dangerous radiation.

ALIEN PROBES

Mora Pol doesn't know that Odo is sentient when first studying him. In revenge, Odo changes into a tentacle and slaps him (DS9 / "The Begotten").

the hot, acidic pores of rock inside volcanoes, or deep underground in icy Antarctica. If bacteria from the *Aquifex* genus thrive in Yellowstone National Park's boiling hot springs, which can reach temperatures of 205°F (96°C), it's not so difficult to imagine that alien species might be able to grow in similarly harsh conditions. As we learn more about these newly discovered life-forms here on Earth, our sense of the possible shapes life might take in our galaxy expands.

ALIENS ABOUND

Over the course of their travels, the members of Starfleet come into contact with hundreds of alien races across the galaxy, from familiar-looking humanoid aliens and cybernetic beings to sentient bloblike entities and noncorporeal aliens.

VULCANS The Vulcans were the first aliens to make contact with Earth and, according to some theories, share a distant ancestor with the Human race. They are best known for their stoicism and cultural values of logic and knowledge. The Vulcan brain has some inherent telepathic abilities and can read the minds of life-forms with a single touch—known as the Vulcan mind-meld. Vulcans have copper-based blood, which sometimes lends a slight green tinge to their skin.

KLINGONS A proud race from the planet Qo'noS (pronounced kronos), the Klingons value honor and skill in combat. The Klingon Empire is a forceful, and often adversarial, interstellar military power. Though he hails from a long lineage of warriors, Worf (at right) becomes the first Klingon to enter and graduate from Starfleet Academy, rising to the rank of lieutenant commander and serving as the Federation's ambassador to Qo'noS.

ANDORIANS **Hailing from icy moon Andoria, blue-skinned Andorians are a militaristic race valuing honesty and conviction, seen in *Ushaan,* an honor code tradition demanding a fight to the death.**

BORG **A pseudo-species of cyborgs, the Borg exist not as individuals, but as a hive mind called the Borg Collective. At the helm, the Borg Queen forcibly assimilates life-forms to the Collective.**

ROMULANS **Close biological relatives of Vulcans, Romulans split off in favor of a more violent, militaristic lifestyle. Though similar to Vulcans, many Romulans have distinct forehead ridges.**

FERENGI **The Ferengi have four-lobed brains impervious to telepathic species, and they are known to have life spans in excess of one hundred Earth years.**

HORTAS **Silicon-based life-forms from Janus VI, Hortas feed exclusively on rock, leaving behind tunnels. Every 50,000 years, all Hortas die except the mother Horta, who watches over the eggs.**

TALOSIANS **An ancient, intelligent race native to Talos IV, Talosians have the ability to create extremely clear mental illusions. After their planet is nearly destroyed, survivors subsist on these illusions.**

A HABITABLE PLANET IN THE NIGHT SKY

Look for Tau Ceti, the host star for a potential habitable planet, within the constellation Cetus.

STARGAZING TIPS

BEST VIEWING SPOTS: From south of 70 degrees latitude in the Northern Hemisphere, and all of the Southern Hemisphere

BEST TIME TO SEE IT: Evening, October to January

BASIC TIPS: Tau Ceti is a faint yellow magnitude 3.0 star, visible to unaided eyes from places with minimal light pollution. This unassuming sunlike star is located less than 12 light-years from Earth, and is part of one of the largest constellations in the sky: Cetus, the Whale. Despite its large size, this stellar character does not contain any bright stars, so tracing its pattern will be more challenging under light-polluted skies, but relatively easy from the countryside. Sky watchers can find the sea creature's head pointing toward its neighboring, much brighter constellation, Taurus.

HOW TO FIND IT

1. In late fall or early winter, gaze toward the eastern horizon at night and look for the bright orange star Aldebaran in the constellation Taurus rising. Aldebaran marks the base of a giant V-shaped group of stars called the Hyades cluster, which acts as a directional guide for locating the Cetus constellation next door.

2. Draw an imaginary arc from Aldebaran and the Hyades cluster directly to the next brightest star called Menkar. It should be about two fist widths—held at arm's length—from the point of the V-shaped cluster to Menkar. Note: *Star Trek* fans will probably recognize this star by its more official name Alpha Ceti. The star hosts the planet Ceti Alpha V on which Captain Kirk originally exiles the villain Khan and his crew. But after neighboring world Ceti Alpha VI is destroyed, the planet is rendered uninhabitable.

3. Continue to follow the arc from Aldebaran to Menkar until you arrive at Tau Ceti, the next brightest star.

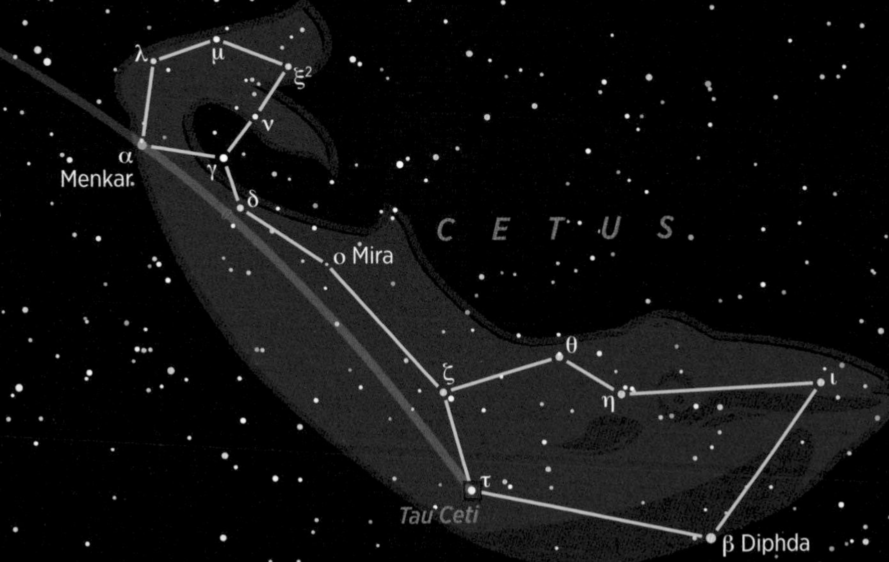
Hyades
Aldebaran
λ
μ
ξ2
ν
α
Menkar
γ
δ
CETUS
o Mira
θ
ζ
η
ι
τ
Tau Ceti
β Diphda

NX-01

SAILING TO THE STARS

STARS FORM FROM GIANT CLOUDS OF DUST AND GAS, FROM PROTOSTAR TO SINGLE AND MULTISTAR SYSTEMS.

CHAPTER THREE

STAR TREK & US

Lighting the *Star Trek* Milky Way are more than 100 billion stars, glowing energy spheres around which are the potential for as many systems of planets, moons, comets, and asteroids. Stars are the most widely recognized astronomical objects in the night sky and are a key element in building galaxies. In the late 21st century, the first long-range Starfleet missions launch into the stellar seas. Federation starships continue to warp thousands of light-years to investigate all types of intriguing stars, the radiant objects like markers guiding them through the universe.

But stars aren't just guideposts for explorers: They're often destinations in and of themselves.

Our stellar fascination traces back for millennia, both in *Star Trek* and real life. The mysterious twinkling lights overhead were revered—considered supernatural in some cultures—until modern observation revealed their true nature. In both realms, the science is as captivating as the mystic: A star's evolution, from birth to death, determines what shape it will take and what materials and possibilities it will contain.

A STAR IS BORN

In real life, we know that stars are giant spheres of gas made mostly of the two most abundant elements in our universe: hydrogen and helium. They release enormous amounts of energy in the form of light and heat and have life spans lasting from a hundred million years to more than ten billion.

Stars are born within the clouds of dust scattered throughout most galaxies. Gravitational disturbances, like colliding objects, give rise to knots of material. As these knots gain mass, the swirling gas and dust start collapsing under their own gravitational attraction, and the center heats. Hydrogen and helium nuclei collide and form new nuclei in a process called nuclear fusion, which produces energy and further heats the young star—known at this point as a protostar. As the star evolves, the same reaction produces the energy that counteracts a gravitational collapse into itself, sustaining it for millions or billions of years.

LIFE CYCLE SPECTACULAR

Most of the stars in our universe, including our sun, are what we call main sequence stars. They steadily burn hydrogen into helium, neither expanding nor contracting until their fuel runs out. In smaller and cooler low-mass stars, it can take up to ten billion years for this to occur. Higher-mass stars, on the other hand, burn through their hydrogen fuel more rapidly and bloat into larger and brighter stars that die more quickly. Thus, mass

PREVIOUS PAGES: The Enterprise *NX-01 explores stars in their many life stages, from their origins in cosmic clouds to their growth and ultimate (sometimes explosive) deaths.*

STARS GALORE

The bright cluster of stars shown at the center live in the very active star-forming region R136 around the Tarantula Nebula in the Large Magellanic Cloud, a small neighbor of the Milky Way. The Tarantula Nebula is about 160,000 light-years away, and R136 produces much of the energy that makes it visible.

is the primary driver in a star's life cycle and determines subsequent characteristics like color, brightness, surface temperatures, and size.

LOW-MASS STARS

Stars of less than ten solar masses evolve over billions of years, swelling, cooling, and gradually fading from view.

RED GIANTS Low-mass main sequence stars eventually become red giants. When a star's core exhausts its hydrogen fuel reservoir and gravity crushes it, its outer layers swell for a few hundred million years, expanding up to a hundredfold into a brighter and cooler star.

PLANETARY NEBULAE When a red giant expands beyond the reach of its core gravity, it begins to shed outer layers, ejecting a shell of gas that expands into what is known as a planetary nebula. The superhot core at the center emits intense ultraviolet radiation that ionizes the surrounding gas, causing it to glow.

BORN IN NEBULAE
Cold gas clouds collapse and matter accumulates on a protostar.

HIGH-MASS STARS

MAIN SEQUENCE

Composition is usually 98% hydrogen and helium.

- 10–100 solar masses
- 90% of life span
- Spica, Theta Orionis C

GIANT/SUPERGIANT

Massive stars are capable of producing heavier elements, such as iron, through fusion.

- significant loss of mass
- 10% of life span
- Betelgeuse, Rigel

LOW-MASS STARS

MAIN SEQUENCE

Composition is usually 98% hydrogen and helium.

- 0.08–8 solar masses
- 90% of life span
- sun, Altair

RED GIANT

Expending hydrogen in their cores, these stars extend their outer layers and can grow to > 100 times their main sequence size.

- 99% of original mass
- 10% of life span
- Aldebaran, Arcturus

WHITE AND BLACK DWARFS As the nuclear fuel fizzles out in a star's core, the naked carbon center cools and collapses under its own weight until it is no bigger than Earth but still has the mass of the sun. Its matter is a million times denser than water—so dense that a teaspoon of it would weigh five tons. As this tiny white dwarf continues to contract under gravitational force, it cools and fades over billions of years to become a black dwarf.

HIGH-MASS STARS

When a star's mass is greater than ten solar masses, it follows a different—and rather more dramatic—path.

SUPERGIANTS When a high-mass star runs out of fuel, its heat and energy start to burn

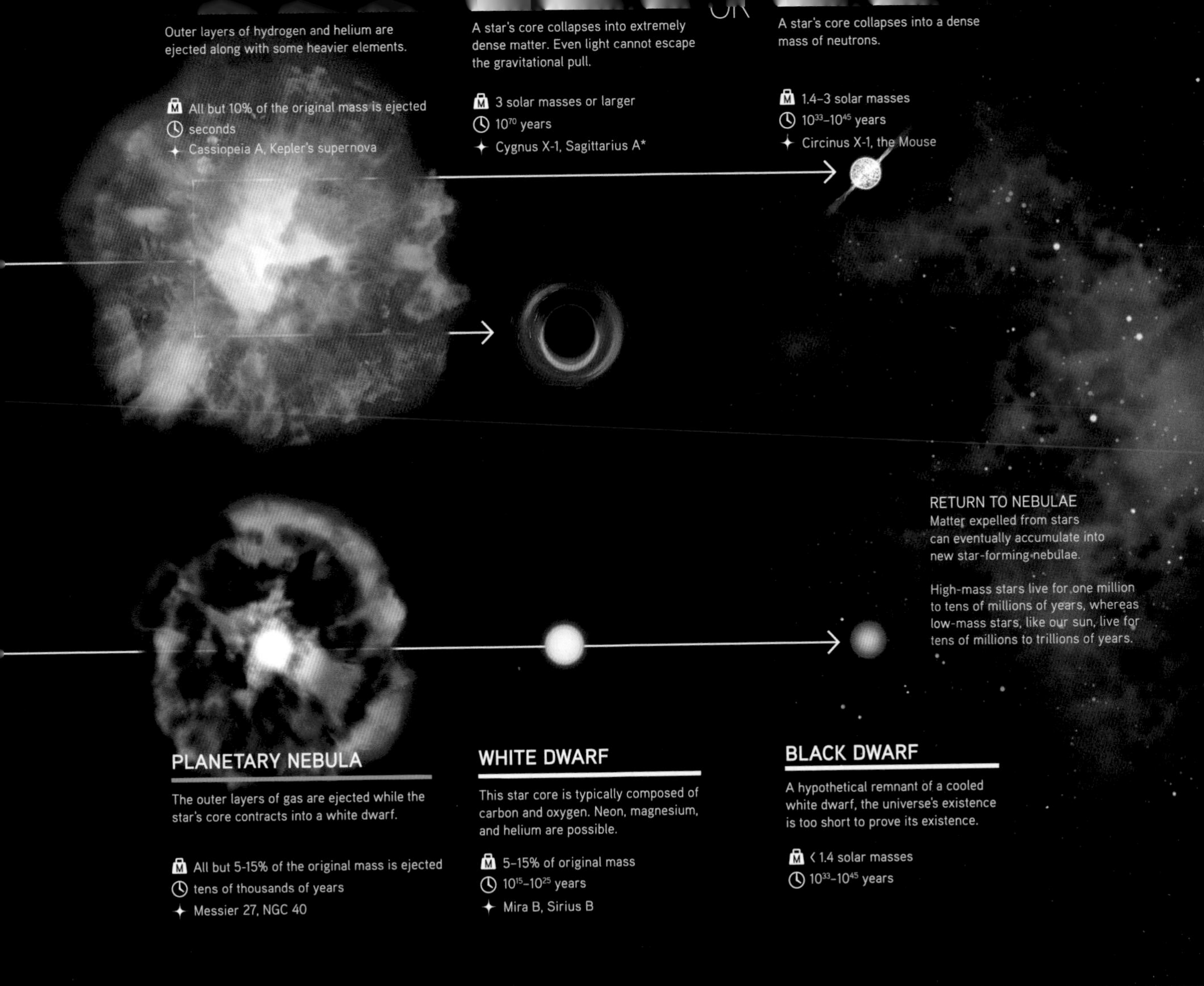

through heavier elements than a lower-mass star can. Eventually, it enters the supergiant phase. The diameters of these stellar monsters can measure more than 600 million miles (966 million km), or six and a half times the distance from the sun to Earth.

SUPERNOVAE A supergiant will burn through its fuel within a few million years and collapse under its own weight. Instead of becoming a planetary nebula, the crushing of its core triggers a massive explosion called a supernova. This cataclysmic death can be so bright that it outshines billions of stars within its home galaxy before fading away.

NEUTRON STARS AND BLACK HOLES Star cores that survive gravitational collapse and are less than three times the sun's mass become a dense mass of neutrons called neutron stars. More massive star remnants also collapse under their own weight, but they continue to shrink until gravity crushes the core into an infinitely dense point called a black hole.

"Captain's log, supplementary. Our investigation of Gamma Triangulі VI has suddenly turned into a nightmare."

—JAMES T. KIRK

MAIN SEQUENCE STARS IN *STAR TREK*

TOS / "THE APPLE"
SEASON 2 / 1967

IN THIS EPISODE: When the *U.S.S. Enterprise* crew discovers an idyllic landscape on planet Gamma Trianguli VI, they are impressed by its lush forests and Earthlike temperature. Though at first it seems like a fertile oasis, it proves full of dangerous shooting pod plants and malevolent hazards that disrupt their antimatter pods and electromagnetic frequencies. When the crew becomes trapped on Gamma Trianguli VI, they discover that a machine deep underground is what controls the planet, keeping its inhabitants healthy and young.

STARFLEET IS ALWAYS LOOKING FOR NEW WORLDS where life (especially humanoid life) can exist. When Gamma Trianguli VI comes into view—the sixth planet in the Gamma Trianguli star system—the *Enterprise* crew sees a planet that looks like a red-tinted Earth: green forests, blue seas, and thick clouds. When they beam down, conditions only seem to get better: fertile soil and blooming plants, an atmosphere that filters out radiation, little variation at the poles, and a steady temperature of 76°F (24°C). Vaal, the machine that powers it, serves to make this planet seem like an oasis—but it doesn't take long for things on Gamma Trianguli VI to get, as Scotty puts it, "a wee bit strange." Vaal doesn't take kindly to its Human invaders, and it quickly turns this paradise into something hellish.

Like most stars in the *Star Trek* universe, Gamma Trianguli's conditions suggest where the star is in its evolution. Its sunlike character places it firmly amid main sequence stars. Much like in real life, *Star Trek* main sequence stars can range from large, hot O- and B-type stars to cooler, smaller K- and M-type stars, the red dwarfs. Red dwarfs are often host to class M planets (the most Earthlike), and thus are of interest when it comes to seeking out alien species and habitable worlds. Not all of *Star Trek*'s main sequence stars are as life-compatible as Gamma Trianguli, but many of them host habitable planets. There is red dwarf Proxima Centauri, only four light-years from Sol, and the Vulcan system's 40 Eridani C.

The Enterprise *approaches Gamma Trianguli VI, a seemingly Terra-like world. With its mild and constant temperature, it is able to support a rich variety of life.*

UNLUCKY REDSHIRTS

In "The Apple" (TOS), four crewmembers wearing red shirts meet an untidy end, one at the hands of a poison pod plant seen here, affirming the term "redshirt" as representing an expendable character.

MAIN SEQUENCE STARS IN OUR UNIVERSE

RED DWARF MOST COMMON TYPE OF STAR / UP TO 5,000°F (2,700°C)

All stars spend most of their lives on the "main sequence" of their evolution. From cool red dwarfs to hot blue stars, these stellar objects are what crowd the cosmos. Some are a tenth of the sun's mass, whereas others can be 100 times that.

THERE'S ORDER TO THE SEEMING CHAOS OF OUR star system, as charted in the Hertzsprung-Russell (H-R) diagram. When we plot the luminosity of stars (the total energy radiated by the star per second) against their surface temperature, we find that the stars aren't randomly distributed, but rather are restricted to a few defined regions. Like the periodic table of elements, the H-R diagram reveals commonalities in the characteristics of stars. But unlike the elements' fixed places in the periodic table, a star's position on the diagram changes as its physical properties evolve. The H-R diagram is more of a visual plot of stellar evolution, used to study what we observe from afar and what Starfleet studies up close.

POPULAR COSMIC PLAYERS

Main sequence stars dominate the sky because they spend around 90 percent of their lives there. They occupy a diagonal band that runs from the upper left corner of the H-R diagram (where hot, luminous stars are charted) to the lower right corner (where cool, dim stars live). These stars have transitioned from protostars to stars that have reached hydrostatic equilibrium—in other words, the stability created when the hydrogen fusion happening in their core produces an outward pressure that balances with the inward pressure of gravitational collapse.

LONG LIVE THE DWARFS

Where a star fits on the H-R diagram's band is a question of mass, the sun being the scale against which all other stars are measured. The sun equals one solar luminosity and one solar mass—a G spectral class star with a surface temperature of about 10,000°F (5500°C). The cooler and smaller a main sequence star is, the longer it lives on the main sequence. Stars like Spica, a rare blue dwarf 18 times bigger than our sun, are destined to burn through their fuel and leave

ARE WE THERE YET?

BEAM ME UP!

As fun as it might look to be "beamed up" from planet to ship, quantum mechanics makes the potential reality seem downright terrifying. We've teleported quantum information between photons over more than ten miles (16 km), but that isn't teleportation in the *Star Trek* sense. The photons in one place are destroyed and remade in the other place. Even if teleportation were possible, it would definitely be dangerous.

Light-Years Away

TOS / "The Return of the Archons"

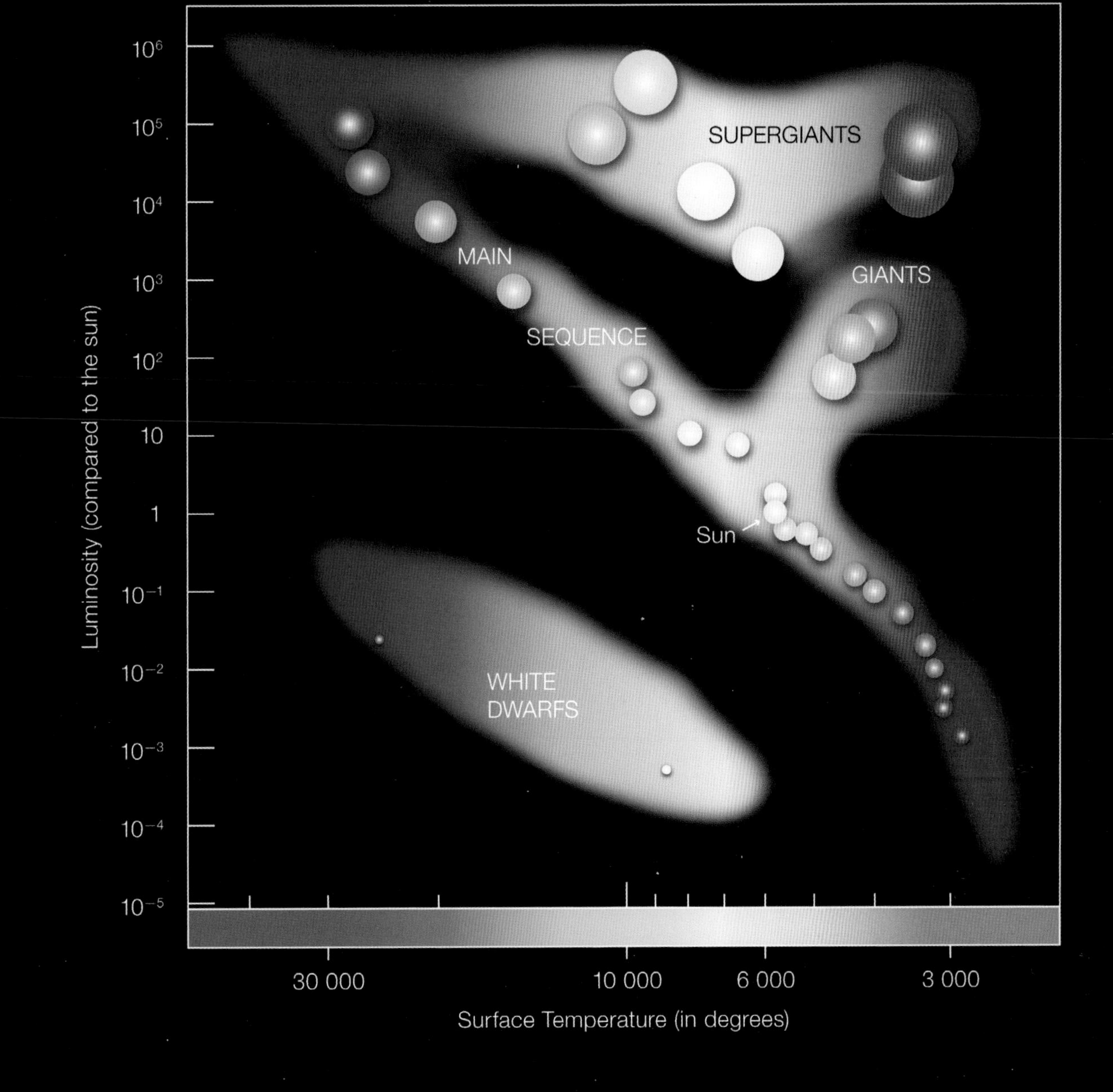

the main sequence quickly. *Star Trek* imagines counterparts for real-life main sequence classes in the form of white F-type stars like Orellius Minor; yellow G-type stars like Calindra; and red M-type stars like 40 Eridani C. 

The Hertzsprung-Russell (H-R) diagram plots stars according to luminosity and temperature (measured here in kelvins). A star's position shows where it is in its life cycle. Stars spend most of their lives on the main sequence.

NON-FEDERATION SHIPS

Ships from outside the Federation are often seen in conflict, from border squabbles with the Klingon and Cardassian armies to all-out attacks from the Ferengi and the Romulans. Often, the distinct styles and attack formations of these ships mark them as alien forces and reflect the intention of their crews.

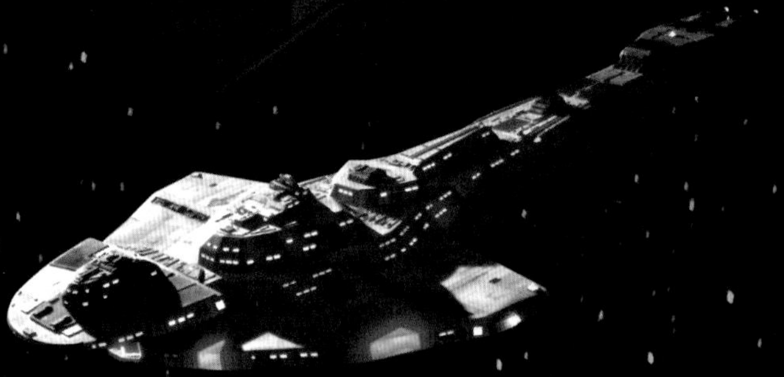

CARDASSIAN WARSHIP **_Galor_-class warships are the main vessels used by the Cardassian military. Armed with a multitude of phaser arrays, these ships are often operated in squadrons of three in battle.**

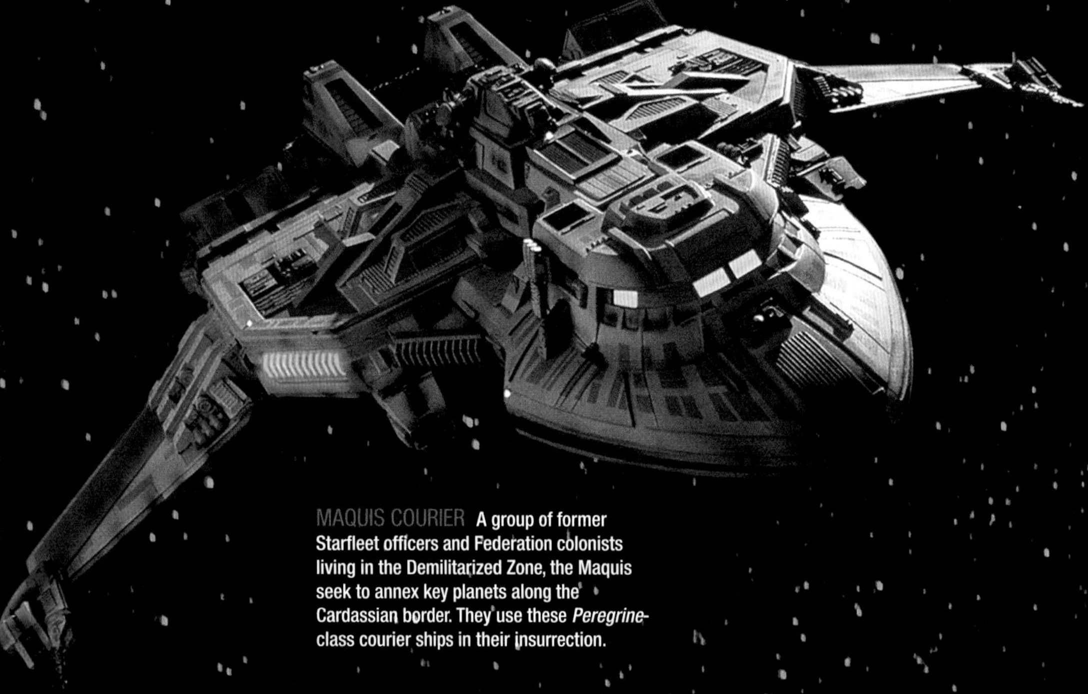

MAQUIS COURIER **A group of former Starfleet officers and Federation colonists living in the Demilitarized Zone, the Maquis seek to annex key planets along the Cardassian border. They use these _Peregrine_-class courier ships in their insurrection.**

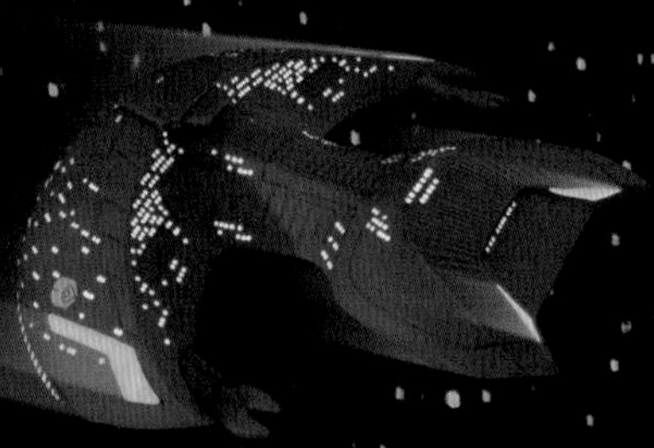

FERENGI MARAUDER **The Ferengi use _D'Kora_-class Marauders for both transportation and attack, depending on circumstances. Marauders are equipped with various powerful directed-energy weapons, as well as torpedo and missile launchers.**

KLINGON BIRD-OF-PREY One of the most versatile ships in the Klingon fleet, the Bird-of-Prey is used as a scout, raider, patrol ship, and cruiser. As its name suggests, this ship is modeled after a predatory bird on the attack, and contains powerful weaponry to use in coordinated attacks.

KLINGON D7 CRUISER A backbone of the Klingon Defense Force, the D7 battle cruiser is a massive ship, some 750 feet (230 m) long. During a brief alliance between the Klingons and the Romulans in the 2260s, the D7 design is shared between the empires, with the Romulans adding cloaking that could make the ships nearly undetectable.

KAZON CARRIER These Kazon Collective warships are armed with particle-beam weapons and plasma torpedoes. Though clumsy in battle, their powerful weapons cache make them formidable opponents.

DWARF STAR IN THE NIGHT SKY

Look for Sirius, Alpha Canis Majoris, a main sequence dwarf, within the constellation Canis Major.

STARGAZING TIPS

BEST VIEWING SPOTS: Across temperate latitudes and south in the Northern Hemisphere in winter, and summer in most of the Southern Hemisphere

BEST TIME TO SEE IT: Early evening, December to April

BASIC TIPS: Only 8.6 light-years away and shining at -1.46 magnitude, Sirius is the brightest star in the sky and is our fifth closest stellar neighbor. This bright white star, nicknamed the Dog Star, dominates the constellation Canis Major. Because Sirius hangs fairly low in the sky from mid-northern latitudes, it appears to twinkle or shimmer. This flickering effect is due to the razor-thin starlight piercing Earth's atmosphere and traveling through a thick column of air that scatters its beam of light before reaching our eyes.

HOW TO FIND IT

1. Face the southern sky in midwinter and look for the constellation Orion, the Hunter, and its distinctive belt of three bright stars. Draw an imaginary line across the belt stars from right to left.

2. Sweep farther east or to the left of Orion's belt with the imaginary line pointing to the next brightest star in this part of the sky, Sirius. This line will be about eight times the width of Orion's belt.

3. Sirius is not alone; it is a double star that has a tiny 8th magnitude companion called the Pup. No bigger than Earth, the Pup is the exposed white core of a dead star. Classified as a white dwarf, it is a challenging target for small backyard telescopes.

Orion
θ
μ Isida
Muliphen γ
ι
α Sirius
β Murzim
Yeji ν2
π
CANIS MAJOR
Menkelb Posterior
ο2
ο1 Menkelb Prior
Wezen δ
σ
ε Adara
η Aludra

"This red giant might be a red herring . . . You don't know that the weapon is there."

—CHARLES "TRIP" TUCKER III

RED GIANTS IN *STAR TREK*

ENT / "STRATAGEM"
SEASON 3 / 2004

IN THIS EPISODE: The *U.S.S. Enterprise* NX-01 captures Degra, the mastermind behind the Xindi superweapon project, and Archer tries to trick him into revealing the location of the final weapon. They suspect that the weapon is being held on a red giant star, but the fact that there are seven spread over about 40 light-years means they're desperate to find out which one it is. With memory wiping, a days-long simulation in which Captain Archer pretends to be Degra's ally, and some out-and-out lies, they finally trick their prisoner into revealing what they hope is the weapon's final location.

WHEN CAPTAIN ARCHER IS CALLED BACK TO THE bridge after interrogating their Xindi prisoner, he looks to the viewscreen and is confronted by a glowing red orb just ahead of them. This fiery vision is only a trick they've concocted to get Degra to reveal the location of the superweapon, but that doesn't make this red giant's dark, simmering spots and flares look any less forbidding. In the *Star Trek* universe, red giants are what main sequence stars become when they grow old. These stars transition into giants when they run out of fuel in their cores and start to fuse heavier and heavier elements. They release more energy, swelling and turning red as these dying suns start to enter their death throes. Very large and fairly cool, these stars are not uncommon in the *Star Trek* universe. The Bopak system's central star is a red giant (DS9 / "Hippocratic Oath"). Three Ferengi decide to use a pair of red giants to try to create a geodesic fold, which they hope will allow them to steal Seven of Nine's nanoprobes (VGR / "Inside Man").

These stars may be on their way out, but their shifting status only makes them more interesting as sites of scientific study. A scientist named Doctor Timicin tries to save a dying sun of his homeworld, Kaelon II, by using photon torpedoes to try to regulate its increasing temperature (TNG / "Half a Life"). A red giant becomes an obstacle in the Antarian Trans-stellar Rally, forcing Tom Paris and B'Elanna Torres to adjust for the photonic interference it causes (VGR / "Drive").

Archer and the Enterprise *crew wipe three Xindis' memories, and then leave this test site for Azati Prime, where they hope to find the Xindi superweapon.*

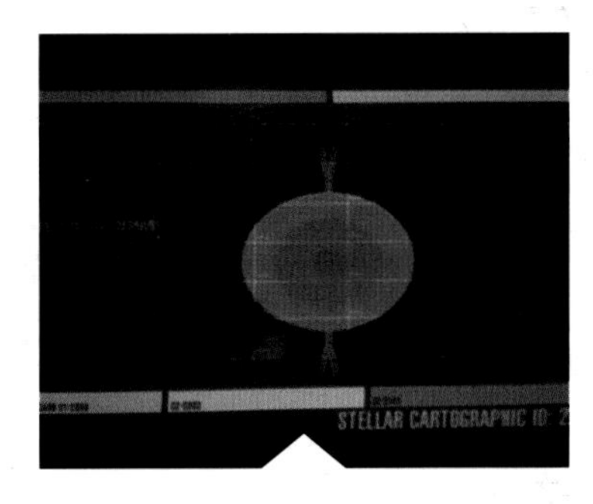

LEANING ON A GIANT

The Pathfinder Project considers utilizing a red giant to rescue the U.S.S. Voyager *from Delta Quadrant, but radiation levels pose too big a risk (VGR / "Inside Man").*

RED GIANTS IN OUR UNIVERSE

DIAMETER 62 MILLION TO 621 MILLION MILES (100M TO 1B KM)
SURFACE TEMP UP TO 5800°F (3200°C)

When low- to medium-mass stars expand and outgrow the main sequence, they enter the realm of giants. As a star burns through its fuel and compacts, the outer atmosphere heats up and expands, dramatically increasing the star's size and multiplying its luminosity.

HOW LONG A STAR LIVES DEPENDS ON ITS MASS. Higher-mass stars may have more material, but they burn through it more quickly because greater gravitational forces increase their temperature and density. To put it into perspective, the sun, a typical yellow dwarf star, will spend about ten billion years on the main sequence. A red dwarf half as massive as the sun can stick around for 80 to 100 billion years—far longer than the age of the universe. But a star ten times more massive than the sun will only last for about 20 million years. Stars on the main sequence are all different sizes, with solar masses at a tenth of the sun's mass to more than 100 times that. No matter its initial mass, stars are destined to leave the main sequence and begin the spectacular process of dying, which often begins when they become giants.

FIERY TITANS

Main sequence stars maintain a state of equilibrium—until they start to run out of fuel. When a main sequence star depletes its hydrogen supply, its equilibrium falters and the pressure in its core starts to ebb. The outer hydrogen burning shell creates an outward push, and the star expands into a giant. Its helium core, by contrast, shrinks down to about a third of its original size.

Once a star has left the main sequence and becomes a red giant, its outer temperature cools dramatically to between 4000 and 5800°F (between 2200 and 3200°C), a little more than half as hot as the sun. This temperature reduction causes these stars to shine in the redder part of the spectrum, although they often appear somewhat orange. A star like Aldebaran has a mass that's only 50 percent greater than the sun's, but it shines with over 400 times its luminosity because of its swollen radius.

A STAR IN OLD AGE

These luminous stars will eventually run out of fuel. When that happens, the core shrinks again and moves on to its final stage of dying. Eventually, low- to

ARE WE THERE YET?

PHOTON TORPEDOES

Photon torpedoes are used for all sorts of things in *Star Trek:* Mostly though, they're used to blow things up. They rely on matter and antimatter, separated by a force field that prohibits any untimely explosions. Our real-life prospect for such weapons is problematic. Antimatter is hard to find in nature, and creating it is difficult and costly. Besides that, it needs to be contained in ways we haven't yet mastered.

Light-Years Away

VGR / "Drive" / Season 7 / 2000

Red supergiant Antares, also known as Alpha Scorpii, shines through a star field 500 light-years from the Earth. At this point in Antares's life, almost all of its hydrogen content has been burned. Only the surrounding hydrogen and central helium core remain, creating a surface temperature of about 5120°F (2830°C).

medium-mass giants will become white dwarfs—glowing white orbs lighting up their cloudlike planetary nebulae.

The sun, too, will gradually become a red giant, bloating and expanding until it's stretched well past the orbit of Venus. All the inner planets will be engulfed. After that, the sun will eject a shimmering nebula and leave behind a compact white dwarf. This process happens so gradually that we have half a billion years to worry about what it means for our survival—and how to ensure our longevity as the sun brightens on its way to becoming a red giant and as our planet's surface gets too hot for liquid water.

BLAST FROM THE PAST

The Think Tank was proud of the fact that they were able to reignite the red giants of the Zai cluster (VGR / "Think Tank").

A RED GIANT SUN

An artist's rendering gives an Earth view of the dying sun in the midst of its red giant phase, some five billion years in the future. This end-of-life stage is expected to last 700 million years, during which intense heat and radiation will sterilize Earth and boil away its oceans.

RED GIANT STAR IN THE NIGHT SKY

Look for Arcturus, on Alpha Bootis, a bright red giant within the constellation Boötes.

STARGAZING TIPS

BEST VIEWING SPOTS: The Northern Hemisphere in late spring and early summer, and autumn in the Southern Hemisphere

BEST TIME TO SEE IT: Early to late evenings, May to August

BASIC TIPS: Situated 37 light-years away, Arcturus is considered the closest giant star to Earth. At 25 times the diameter of our sun, it radiates about 100 times more light, making it appear as a very bright -0.3 magnitude orange-hued star that dominates the late spring skies. Arcturus pins down the left foot of the constellation Boötes, the Herdsman from Greek mythology. However, to many of today's sky watchers, the ancient star pattern looks more like a giant kite or an ice-cream cone in the sky.

HOW TO FIND IT

1. In temperate northern latitudes in late nights of early summer, look for the familiar star pattern of the Big Dipper high in the northwest, standing on its bowl. The handle of the dipper is composed of three stars that curve downward.

2. Follow the natural curvature of the Big Dipper's handle and continue out toward the near overhead sky, following the arc until it arrives at the next brightest star: orange-hued Arcturus hanging high in the southern sky.

3. Use binoculars to see Arcturus's soft golden tint. This star is ranked as the fourth brightest in the sky. As an added observing bonus for small backyard telescopes, look for nearby double-star system Izar, which has a beautiful orange color.

Dubhe
Merak
Big Dipper
Megrez
Alioth
Phecda
Mizar
Asellus
Tertius
Asellus Primus θ
κ2
Asellus Secundus ι
Alkaid
Aulad al Thiba λ
β Nekkar
γ Seginus
δ Princeps
BOÖTES
ρ Hemelein Prima
Izar ε
α Arcturus
η Muphrid
τ
ζ

"No earth ship has ever been within ten light-years of a hypergiant. How much closer can we get?"

—JONATHAN ARCHER

SUPERGIANTS IN *STAR TREK*

ENT / "COGENITOR"
SEASON 2 / 2003

IN THIS EPISODE: While studying a huge, rare hypergiant star, the *U.S.S. Enterprise* NX-01 meets and befriends the Vissians, a highly evolved humanoid species. Their captain gives Archer the chance to come aboard one of their sophisticated stratopods and venture into the stars' inner sanctum. But even as the two species get better acquainted, it doesn't take long for trouble to brew. Tucker takes exception to the way the Vissians treat their cogenitors—a third gender treated almost like slaves—causing the new alliance to fray.

THIS HYPERGIANT STAR DWARFS THE ENTERPRISE, surrounding it in a brilliant yellow-gold light. Its center appears so hot that it glows white, becoming ever redder as it rings outward. Even as the ship moves toward it, the hypergiant seems to have no edge. When Captain Archer travels inside it with the Vissian captain, he remarks that he has friends back at home who wouldn't believe what he is witnessing. He notices purple plumes of ionized hydrogen, which the Vissian captain says is generating magnetic currents. Archer knows he's seeing something rare and magnificent; this is the first time one of Starfleet's ships gets close enough to truly research a hypergiant. Though it's over 600 million miles (almost a billion km) in circumference and hypergiants are known to turn into supernovae when they exhaust their stores of energy, Archer remains unconcerned. That's probably because when they visit in 2152, it isn't expected to go supernova for at least another hundred years.

In the *Star Trek* universe, a hypergiant's luminosity is even greater than that of a supergiant. The two are similar in makeup, but hypergiants are even bigger—supposedly one hundred times the mass of Sol and up to a thousand times bigger. Supergiants and hypergiants alike burn fast and bright, making them fleeting beacons worth looking out for—like the blue supergiant star Alnitak Aa, which Kirk points out to Edith Keeler when he travels back in time to 1930s Earth (TOS / "The City on the Edge of Forever").

SHARED TECHNOLOGY

Vissian ships are notable for their advanced photonic weapons. Shortly after this encounter, the Enterprise *is newly equipped with photon torpedoes, suggesting the groups exchanged technology (ENT / "The Expanse").*

The Enterprise *befriends a Vissian ship while exploring a hypergiant star, and the Vissian customs and technological advancements in turn confuse and delight the crew.*

SUPERGIANTS IN OUR UNIVERSE

UP TO A MILLION TIMES BRIGHTER THAN THE SUN / 30 TO 1,000 TIMES MORE MASSIVE THAN THE SUN

Supergiant and hypergiant stars are the true titans of the cosmos. They burn quickly and furiously, swelling well past the size of Jupiter's orbit before experiencing a violent death that lights up the cosmos.

UNLIKE RED GIANTS, ONCE HIGH-MASS STARS start expanding, they begin to fuse helium almost immediately. They're huge and hot enough that they can burn through elements much heavier than helium, like neon, magnesium, silicon, and sulfur.

LARGER THAN LIFE

Supergiants can reach up to 1,500 times the sun's radius, desperately spewing an incredible amount of energy to prevent gravitational collapse. The rate of fusion within a supergiant can vary wildly as different combinations of elements fuse. This variation means that supergiants can range from red to blue and back again.

Red, highly luminous supergiants have spectral types K and M, which means they're relatively cool at the surface, with temperatures measuring less than 7000°F (almost 4000°C). Betelgeuse, visible in the constellation Orion, is a familiar red supergiant that's estimated at 1,000 times the sun's radius and churns out 135,000 times more energy. Periods of slow fusion cause a star to contract and become a rare blue supergiant—hot, large-mass stars with smaller diameters than their red supergiant siblings. Elusive even in the *Star Trek* universe, blue supergiants are smaller than 100 solar radii, have very high luminosities, and tend to be unstable because of how quickly they shed their mass. Like blue supergiant Rigel, they have extreme surface temperatures and burn so ferociously they're bound not to live long.

THE BIGGEST OF THEM ALL

Grander even than supergiants are hypergiants. The true monsters of the universe, hypergiants like the larger star in the binary Eta Carinae—which has about 90 solar masses alone and an incredible 120 total in the binary system—have proportions that boggle the mind. Leviathans like these live fast and die young, burning through their fuel in a fleeting few million years.

ARE WE THERE YET?

SHUTTLES IN SPACE

In *Star Trek*, stratopods enable Starfleet explorers to dive into a star. In real life, we don't yet have anything that compares. There's the Manned Maneuvering Unit, which lets astronauts move around a ship, but it's more of a fancy backpack. Companies like Blue Origin and SpaceX are working on making space tourism a reality, but we may have to wait for Starfleet-like ships to be developed.

Light-Years Away

Blue Origin space vehicle

That brings us to the next phase in a star's evolution: stellar death. The most massive stars cannot lose enough of their mass to become a white dwarf. Instead, they will always end their lives with a spectacular bang.

Red supergiant Betelgeuse and blue supergiant Rigel appear to Earth at the upper left and bottom right of the constellation Orion. Betelgeuse is so evolved that it will go supernova in the near future—at least on the cosmic time scale.

THE SWAN'S SUPERGIANT

Supergiant Deneb (upper left) stands out in a dazzling shot of the night sky. As seen from Earth, Deneb is in the "tail" of the constellation Cygnus, the Swan. It is one of the three stars of the Summer Triangle asterism, or star pattern, seen from the Northern Hemisphere, and its white color shows its surface temperature is 13,000 to 17,500°F (7200 to 9700°C).

SUPERGIANT STAR IN THE NIGHT SKY

Look for Deneb, or Alpha Cygni, a bright supergiant within the constellation Cygnus.

STARGAZING TIPS

BEST VIEWING SPOTS: Most of the Northern Hemisphere and circumpolar north of 45 degrees latitude, and in the Southern Hemisphere from only north of 45 degrees latitude south

BEST TIME TO SEE IT: Late evenings, June to November

BASIC TIPS: Deneb marks the tail of Cygnus, the Swan constellation. The blue supergiant also marks the head of the asterism called the Northern Cross. Although it is one of the brightest stars in the entire sky, shining at magnitude +1.3, Deneb lies an estimated distance of 3,000 light-years from Earth, making it one if the farthest stars visible to the unaided eye. For it to appear so brilliant, it's estimated to be 100,000 times more luminous and to stretch more than 200 times wider than our relatively puny little sun.

HOW TO FIND IT

1. During late nights of August, look for the Summer Triangle riding high in the southern sky, which will be nearly overhead for temperate northern latitude locations. This large asterism, or stellar pattern, is easy to spot because of its standout brilliant trio of stars, Vega, Altair, and Deneb.

2. Vega, Altair, and Deneb mark the corners of the Summer Triangle. The faintest of the three stars is Deneb, located toward the east.

3. Scan the mythical figure of the swan, from its long neck to its tail, with binoculars to see the Milky Way band and its countless star clouds running through the constellation. For small telescopes, added cosmic attractions in Cygnus include the pretty amber and blue double star Albireo, 380 light-years away and marking the swan's beak or the foot of the cross. The open star cluster M29 near the cross's center star is easy to spot with binoculars and shows well in telescopes.

κ
ι
ο¹
Deneb α
Ruc δ
C Y G N U S
ν
Vega
γ Sadr
η
ε Gienah
Summer Triangle
ζ
φ
β¹ Albireo
Altair

"You are beautiful. More beautiful than any dream of beauty I've ever known."

—SPOCK

SUPERNOVAE IN *STAR TREK*

TOS / "ALL OUR YESTERDAYS"
SEASON 3 / 1969

IN THIS EPISODE: McCoy, Spock, and Kirk plan to visit the planet Sarpeidon to urge inhabitants to vacate before their sun, Beta Niobe, goes supernova. What they find is a lack of inhabitants—just a strange librarian and a vast library filled with the planet's seemingly infinite places and pasts. When the three of them get trapped in some of those pasts and Spock finds himself falling in love with an exiled woman, McCoy and Kirk must convince him to return to their present before the supernova explosion blows them and Sarpeidon into oblivion.

KIRK, MCCOY, AND SPOCK ARRIVE BACK ON THE U.S.S. Enterprise just in time to watch the nearby star burst into a supernova (or, as Kirk calls it, "go nova") as they hastily beat a retreat at maximum warp. What looks at first like a small, red orb pulsing in space behind them appears to draw in on itself, shrinking to a trembling yellow shimmer before blasting outward, bathing the ship in a blue-white light.

A well-documented supernova in both *Star Trek* and our universe is the Crab Nebula. When it exploded in 1054, it was marveled at and written about by stargazers all over Earth. Explosions like the one that almost wipes out the *Enterprise* can emit an electromagnetic pulse that scrambles computer systems. This threat is what leads the technologically advanced Bynars to take over the *Enterprise*-D in the Beta Magellan system, thus saving themselves from extinction (TNG / "11001001").

But not everyone manages to escape a supernova's devastation. Even the incredibly advanced and densely populated Tkon Empire wasn't able to survive it when their central star went supernova (TNG / "The Last Outpost"). In the 24th century, Spock pulls an amazing feat when he figures out how to contain a supernova's destruction: He uses red matter to create a black hole that consumes it, but not before it destroys the planet Romulus (*Star Trek* [2009]). Luckily for civilization, supernovae only tend to occur in the Milky Way a few times a century. By the 24th century, only three Starfleet ships have witnessed one up close.

SHAKESPEAREAN CONNECTION

The title "All Our Yesterdays" (TOS) comes from Shakespeare's play* Macbeth*: "And all our yesterdays have lighted fools / The way to dusty death."

Kirk, Spock, and McCoy barely make it back to the Enterprise *in time to escape Beta Niobe's supernova explosion, which engulfs doomed planet Sarpeidon.*

OCCURS ONCE EVERY 50 YEARS IN THE MILKY WAY / SHOOTS MATTER INTO SPACE AT UP TO 25,000 MPH (40,000 KM/H)

A high-mass star's life ends with a bang instead of a whimper, exploding in the form of a supernova. A star's violent death can give birth to a range of fascinating cosmic offspring, all interesting memorials to their parent stars' pasts.

WHEN HIGH-MASS STARS HAVE BURNED THROUGH all of the lighter elements, they finally get to iron. Iron doesn't release energy when fused, but rather absorbs it. A star cannot sustain its fusion when it tries to fuse a largely iron core, causing its core temperature to drop, letting gravity take over, and triggering the biggest explosion in space.

A BRILLIANT END

There are three types of supernovae that result from high-mass core collapse—Types Ib, Ic, and Type II—all three of which collapse under the weight of their own gravity, an event that happens in a matter of seconds. Iron nuclei are squeezed together, causing the core to bounce off its outer layers. The result is a catastrophic explosion—a brilliant display that briefly outshines the galaxy, radiating as much energy as the sun is expected to emit in its entire life span.

Supernovae are pretty rare in our galaxy, but we've been able to catch them in galaxies outside of our own. Astronomers on Earth spotted a supernova in 1572 that was so bright that it could be seen in daylight despite the fact that it was 5,000 light-years away.

So what would it be like to witness a real-life supernova from the vantage point of a nearby planet? If a star closer than about 100 light-years away went supernova, we would experience an intense blast of heat and light. Our planet would be blasted with radiation equal to that of every nuclear weapon on Earth for days.

FROM DEATH, LIFE

What astronomers are actually seeing when they look at a supernova event is a massive shock wave moving at supersonic speeds through the destroyed outer layers of the star. This hot shell, called a supernova remnant, glows and expands for thousands of years and eventually thins out enough to reveal a neutron star, or even a black hole.

ARE WE THERE YET?

TIME TRAVEL

***Star Trek* explorers travel back and forth through time; in real life we may be able to jump forward, but we definitely cannot go back. Time dilation could allow us to jump forward by traveling close to the speed of light, returning home to find much more time has passed on Earth than the traveler has actually experienced. The trick is you'd have to be traveling very, VERY fast—faster than currently possible.**

Light-Years Away

TOS / "All Our Yesterdays"

Created from visible wavelength and x-ray data, this image of nebula SNR 0509-67.5 uses a pink shell to show the ambient gas being shocked by the expanding blast wave from the Type 1a supernova of which it is a remnant.

The rich material a supernova leaves behind will eventually synthesize into new stars, beginning the stellar cycle anew. Our solar system is thought to have formed from such material, making humans, our planets, and the rest of the systems scattered across the universe the stuff of stardust. 

CARINA NEBULA

A ground-based image of the Carina Nebula captures two of the most massive and luminous stars in the galaxy: Eta Carinae and HD 93129A. Eta Carinae (the brightest star in this image) is brighter than one million suns, and likely has a mass more than 100 times that of the sun.

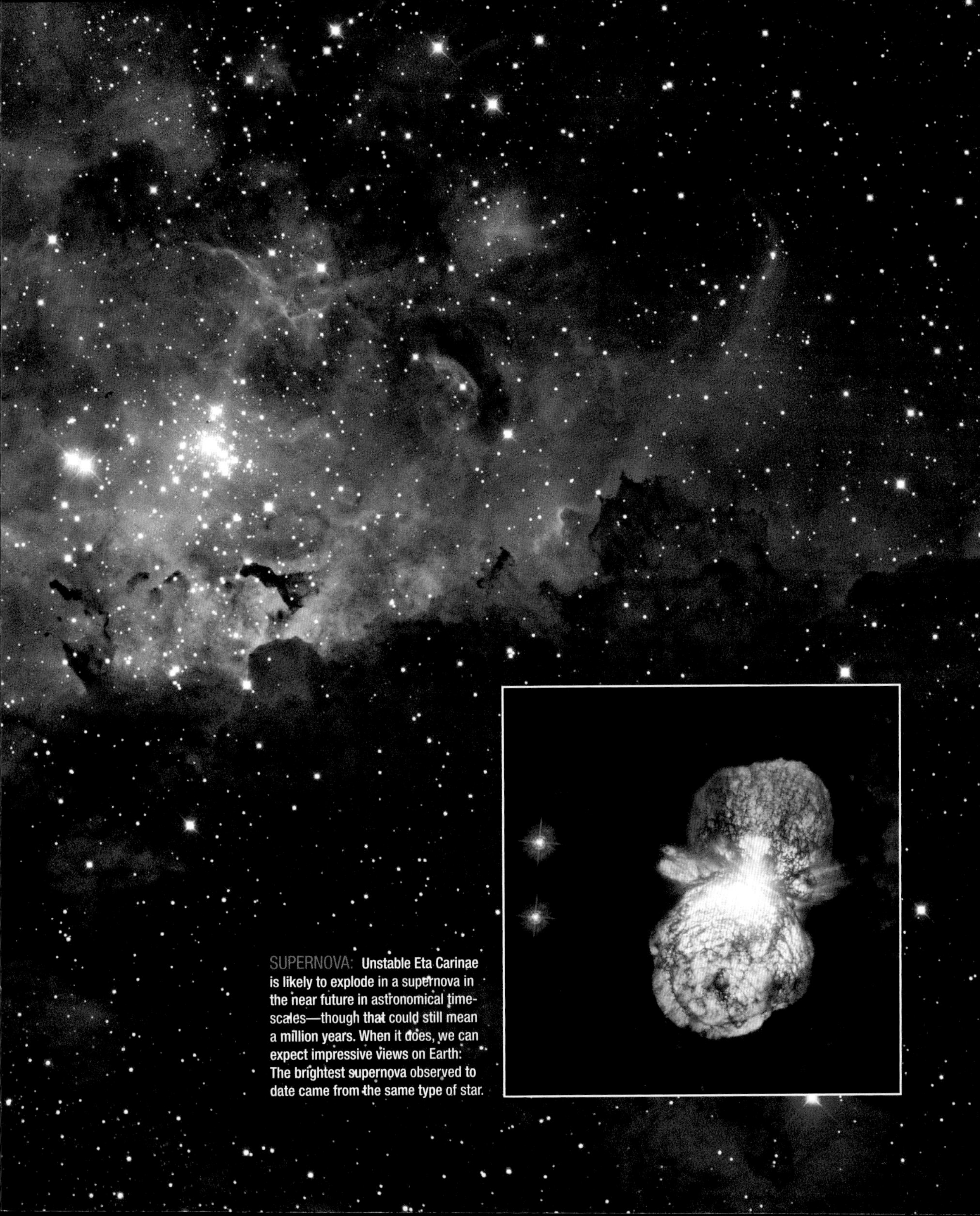

SUPERNOVA: **Unstable Eta Carinae is likely to explode in a supernova in the near future in astronomical timescales—though that could still mean a million years. When it does, we can expect impressive views on Earth: The brightest supernova observed to date came from the same type of star.**

PRE-SUPERNOVA IN THE NIGHT SKY

Look for Antares, or Alpha Scorpii, a star about to go supernova, within the constellation Scorpius.

STARGAZING TIPS

BEST VIEWING SPOTS: From temperate northern latitudes and all southern parts of the Northern Hemisphere, including the southern United States and all of the Southern Hemisphere

BEST TIME TO SEE IT: Late evening, July to October

BASIC TIPS: The brightest star in the Scorpius constellation is Antares. This stellar monster is so large that it dwarfs other stellar giants like Arcturus and Aldebaran. With an estimated diameter 700 times larger than our sun, it may explode as a supernova in the next few million years. When this happens, the 600 light-years distant Antares promises to outshine all other stars in the Milky Way and be visible across the universe.

HOW TO FIND IT

1. Face southward during late summer nights and look for the distinctive Teapot asterism of Sagittarius tipped toward the giant fishhook pattern of stars that mark the curved tail of Scorpius, the Arachnid constellation.

2. Find Antares, the brightest ruddy colored star marking the mythical beast's heart near the far north end of the scorpion's body.

3. Notice Antares's distinctive orange-red hue that reminded ancient astronomers of the planet Mars. The name "Antares" literally means "the Rival of Ares," Ares being the Greek name for the god of war, known as Mars to the Romans. Aim binoculars and telescopes at the stellar giant and spot a tiny fuzzy ball of light—a prominent globular star cluster called M4 located some 7,200 light-years away.

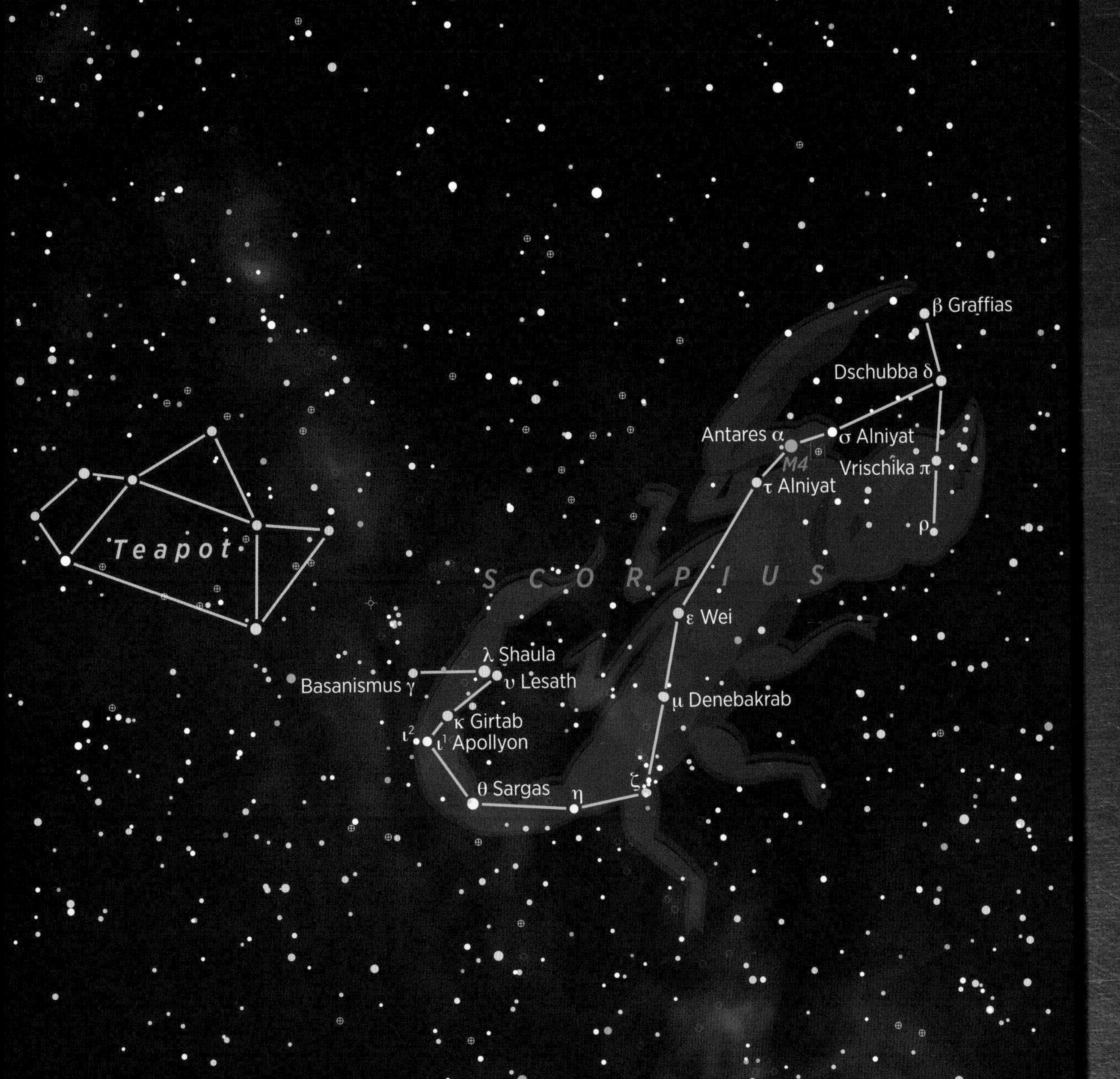
β Graffias
Dschubba δ
Antares α
σ Alniyat
M4
Vrischika π
τ Alniyat
ρ
Teapot
S C O R P I U S
ε Wei
λ Shaula
Basanismus γ
υ Lesath
μ Denebakrab
κ Girtab
ι2
ι1 Apollyon
θ Sargas
η
ζ

"Even if Enterprise *makes it past the black hole without being destroyed, it seems likely the crew won't survive."*

—T'POL

BLACK HOLES AND NEUTRON STARS IN *STAR TREK*

ENT / "SINGULARITY"
SEASON 2 / 2002

IN THIS EPISODE: The crew of the *Enterprise* NX-01 are abuzz with excitement as they near a rare trinary system with a black hole at its center. But as the crew gets closer, their excitement turns into uncharacteristic, aggressive behavior. Crewmembers start fighting over trivial matters, and then they start becoming ill and falling into a coma. It's down to T'Pol, the only crewmember not affected, to figure out how to rescue the crew from their illness and to keep the ship from being destroyed by the black hole.

THE BLACKENED ORB AT THE MIDDLE OF THIS trinary system, burning bright with a pinkish light, appears equal parts beautiful and menacing. The crew marvels at it, its presence establishing black holes definitively for the first time in the *Star Trek* universe (even if, as the crew suggests, at this point Vulcans have already surveyed over 2,000 of them). But what seems thrilling from far away becomes a mortal threat up close. The radiation given off by the trinary system presents a rapidly growing danger; the black hole itself becomes a yawning, dark mouth threatening to eat them alive.

Both in *Star Trek* and real life, black holes are mysterious, alluring, and dangerous. In our universe, we only have well-grounded theories as to what will happen if a spaceship—or a person—is pulled into one. In *Star Trek,* sensors are capable of analyzing black holes *(Star Trek Into Darkness),* and yet they remain both surprising and unpredictable. The Voyager 6 probe vanishes into one only to appear again on the opposite side of the galaxy *(Star Trek: The Motion Picture).* In 2369, the *Enterprise* finds an alien species that seeks out black holes as a nest for their young (TNG / "Timescape").

Black holes are at the heart of some of the most bizarre and cataclysmic incidents that Starfleet has ever had to face. A black hole is so powerful that it actually creates an alternate reality; when Spock's ship and the Romulan *Narada* are sucked into it, they manage to permanently alter both time lines *(Star Trek* [2009]).

The trinary star system glittering in front of the Enterprise *NX-01 contains a class IV black hole—a rare and exciting find that also poses unexpected dangers.*

LOOSE DEFINITIONS

In Star Trek: The Motion Picture, *the Voyager 6 goes through "what used to be called a black hole," suggesting that the definition may have evolved.*

DOT. 1004.527

BLACK HOLES AND NEUTRON STARS IN OUR UNIVERSE

SUPERMASSIVE BLACK HOLES LIKELY LIE AT THE CENTER OF MOST GALAXIES
MILKY WAY HOLDS A FEW HUNDRED MILLION

Post-supernova phenomena come in two flavors: black holes and neutron stars. Both result when a high-mass star's core collapses into a dense, high-mass sphere of matter. Although rapidly spinning neutron stars are the GPS of the galaxy, getting too close to a black hole spells destruction.

THE NAME "BLACK HOLE" BELIES THE FACT THAT it's anything but empty space: It's actually a great amount of matter packed into a tiny volume. Imagine a star ten times more massive than our sun squeezed into a sphere with, theoretically, no size at all, that creates a gravitational field so strong that nothing can escape it.

INESCAPABLE ODDITIES

When the collapsed core of a star is more than three solar masses, a black hole is the inevitable result. Scientists don't actually "see" black holes: They infer their presence by detecting their effect on nearby matter. If a black hole passes through a cloud of interstellar material, it will pull matter inward in a process we call accretion. The same thing occurs if a normal star passes by a black hole—the tidal pull can tear the star apart, accelerating the attracted matter and heating it to a point that it radiates x-rays into space.

As *Star Trek* bends the rules of time as we know it, black holes could upend our understanding of existence. At the moment when a black hole forms, the star's surface nears an imaginary surface called the event horizon—a point of no return where gravity is so strong that escape becomes impossible. Time on the star slows compared to the time kept by faraway observers. Any object approaching the horizon will be seen as moving slower and slower, never appearing to actually cross it.

CITIES AND LIGHTHOUSES

For star cores of less than three solar masses, what remains after a supernova explosion is a neutron star—an extremely dense, metropolis-size remnant formed when a stellar core collapses. These fast-spinning stellar has-beens can have a 12-mile (19-km) diameter and are so dense that just one teaspoon of it would weigh around a billion tons. Supernova 1987A is in the fringes of the Tarantula Nebula and, even though it's in a different galaxy (the Large Magellanic Cloud), is close enough to Earth that its violent death

ARE WE THERE YET?

ENTER THE BLACK

In *Star Trek,* a black hole can destroy space travelers—or it can take them to another reality. We think we know what would happen if we passed near a black hole (in short, a slowing of time before complete annihilation), but we're a long way from proving that passing "through" one is even a possibility. It doesn't look like we'll be taking black holes for a test drive anytime soon.

Light-Years Away

Black hole travel

M82X-2, thought to be a black hole, was later discovered to be the brightest pulsar yet found. Located in the constellation Ursa Major, it pulses approximately every 1.37 seconds.

was visible to the naked eye. In *Star Trek*, astrophysicist Paul Stubbs excitedly boards the *Enterprise*-D to study the way a neutron star's gravitational pull absorbs a nearby red giant's material (TNG / "Evolution").

Some of these neutron stars spin so rapidly that they become pulsars, whose highly magnetized core emits a beam of electromagnetic radiation as the star rotates in short, regular periods. Despite the "pulsar" name, they don't actually pulse—their spin and orientation just make it look that way. Each has a unique and consistent spin period, spinning as fast as 700 times a second. In *Star Trek*, these beacons can do more than shine light. Reginald Barclay is able to aim a tachyon beam at a pulsar and produce a wormhole that allows communication with the *Voyager* (VGR / "Pathfinder").

TURN UP THE VOLUME

Though* Star Trek *doesn't mention the phenomenon, real-life studies suggest that supernovae hum and shake like a giant boom box before they explode.

COMMUNICATION DEVICES

As they travel throughout and beyond the galaxy, members of Starfleet need reliable methods of communication with their crew and other life-forms they might encounter. Many of the Federation's communication devices operate on a subspace frequency, allowing them to transmit data and messages faster than the speed of light (handy when you're sending a message into deep space).

COMBADGE **Badges equipped with communication technology replace handheld communicators in the early 24th century. A single tap activates these compact devices, which can be repurposed as makeshift beacons.**

COMMUNICATOR **When called away from the ship, Starfleet personnel carry these handheld communicators to maintain a connection with the rest of the crew. The signals from these devices also serve as location markers, making them handy for beaming back onto a ship. The Starfleet communicator seen here was used in the 2260s—advances in technology make them smaller and more powerful.**

EARPIECE Earpieces are crucial tools for communications officers like Nyota Uhura. They are vital in her role communicating within the *Enterprise* and with other ships and planets.

VIEWSCREEN A prominent feature in starships, space stations, and other planetary facilities, viewscreens are used for everything from displaying the area in front of a ship and information about other nearby vessels to calling up data and ship-to-ship communication. When in range, visual contact can be made between ships, allowing for face-to-face communication.

UNIVERSAL TRANSLATOR Made possible by a set of ideas and concepts most intelligent life-forms share, the universal translator compares brain wave frequencies, translating ideas into voice and language listeners understand.

BLACK HOLE IN THE NIGHT SKY

Look for V4641 Sagittarii, or V4641 Sgr, a dense remnant of a star, within the constellation Sagittarius.

STARGAZING TIPS

BEST VIEWING SPOTS: From temperate northern latitudes and all southern parts of the Northern Hemisphere, including the southern United States and all of the Southern Hemisphere

BEST TIME TO SEE IT: Late evening, July to October

BASIC TIPS: Astronomers first spotted the black hole V4641 Sagittarii in 1999, when it suddenly had a violent outburst of energy. Located 24,000 light-years from Earth in the constellation Sagittarius, the Archer, the black hole has a companion star off which it feeds. The hapless star's gas is being gravitationally pulled into a disk around the black hole, which then intermittently ignites and produces powerful outbursts of x-rays that Earth's telescopes can detect. Because the black hole rarely produces outbursts that allow it to shine bright enough to be visually seen through backyard telescopes, the cool factor is simply knowing where this cosmic predator is hiding in our sky.

HOW TO FIND IT

1. During dark late-night skies in August or September, look across the southern half of the Northern Hemisphere for the three bright stars that form the Summer Triangle, which will hang nearly overhead in the southeast sky.

2. Draw an imaginary line from Deneb through Altair and continue for the same distance toward the southern sky until arriving at the distinctive pattern of stars called the Teapot, in the constellation Sagittarius. The celestial Teapot asterism is complete with handle, spout, and lid.

3. Star hop from base stars of the Teapot's spout, bright +1.9 magnitude Epsilon Sgr to fainter +2.7 magnitude Delta Sgr and then hop the same distance to the spot where the black hole V4641 Sgr lurks. Sweep this region with binoculars under dark skies to see the Milky Way appearing to rise from the Teapot like steam. Just to the right of its spout is our galaxy's core some 27,000 light-years away, with its own supermassive black hole, partially hidden behind dark star clouds.

Deneb
Vega
Summer Triangle
Altair

Albaldah π
Manubrium ο
μ Polis
Lambda Sgr
λ
V4641 Sgr
Sigma Sgr σ
Tau Sgr τ
φ
Phi Sgr
δ Delta Sgr
Zeta Sgr ζ
γ Gamma Sgr
SAGITTARIUS
Epsilon Sgr ε
η Rabah al Waridah

NCC-74656

CLOUDS AMONG THE STARS

STELLAR BIRTHS AND DEATHS: SHIMMERING APPARITIONS RIPE FOR EXPLORING.

CHAPTER FOUR

STAR TREK & US

Stars and planets may be the leading destinations in *Star Trek*, but the space between is jam-packed with action, too. Not just an empty void, the vast space between the stars is filled with patches of fog called nebulae. These celestial clouds of gas and dust, whether glowing in an array of colors or dark, are sources of scientific wonder in our universe. Amid the space superhighways in *Star Trek*, they play a more dramatic role—murky hideouts for expeditions looking to go stealth; mine sites rich with vital chemicals; and eerie phenomenon worthy of further study.

Radiation from these ghostly shrouds can disable sensory detection on local radars, allowing starships to disappear from, or surprise, enemy vessels. Around 2366, the *U.S.S. Enterprise*-D conceals itself inside the Paulson Nebula to throw off an attacking Borg cube (TNG / "The Best of Both Worlds, Part I"). Nebula theatrics reach a crescendo in *Deep Space Nine* when a major battle between Alpha, Beta, and Dominion forces takes place in the Gamma Quadrant's Omarion Nebula. A fleet of Jem'Hadar fighters—Dominion attack ships—emerges unexpectedly from the nebula to launch a devastating assault on Cardassian and Romulan crafts. The defending fleet is annihilated without hope of retreat, but the *U.S.S. Defiant* fights its way out of the battle zone and back to Deep Space Nine (DS9 / "The Die Is Cast").

Navigating our universe's Milky Way doesn't call for such covert operations, but we are still investigating fascinating nebulae from every possible angle. We understand that gas and microscopic dust permeate the Milky Way's entire disk, and in many regions it concentrates into nebulae. Nearby energy sources illuminate these clumps of matter, lighting them from the inside or outside. When they aren't visible to the eye, radiation emissions of nebulae reveal their location.

BUILDING BLOCKS

Nebulae represent critical stages in stars' life cycles, both at beginning and end. When stars die, their atmospheres—rich in lighter elements like hydrogen and helium—return into space. Heavier elements, such as carbon, silicon, and iron, which are made in the nuclear furnaces of more massive stars, enrich the gases and form molecules of dust that attract more matter. These clumps contain the basic ingredients of stars; they evolve into a kaleidoscope of stars and planet types, and when they die, the process begins anew.

About two dozen nebulae can be spotted with the naked eye from Earth—faint, glowing patches amid a myriad of stars. Ancient astronomers first spotted these hazy-looking phenomena and named them nebulae—the Latin word for "cloud." In the

PREVIOUS PAGES: The Voyager *explores all sorts of phenomena in the distant (and previously uncharted) Delta Quadrant, from star clusters to quasars.*

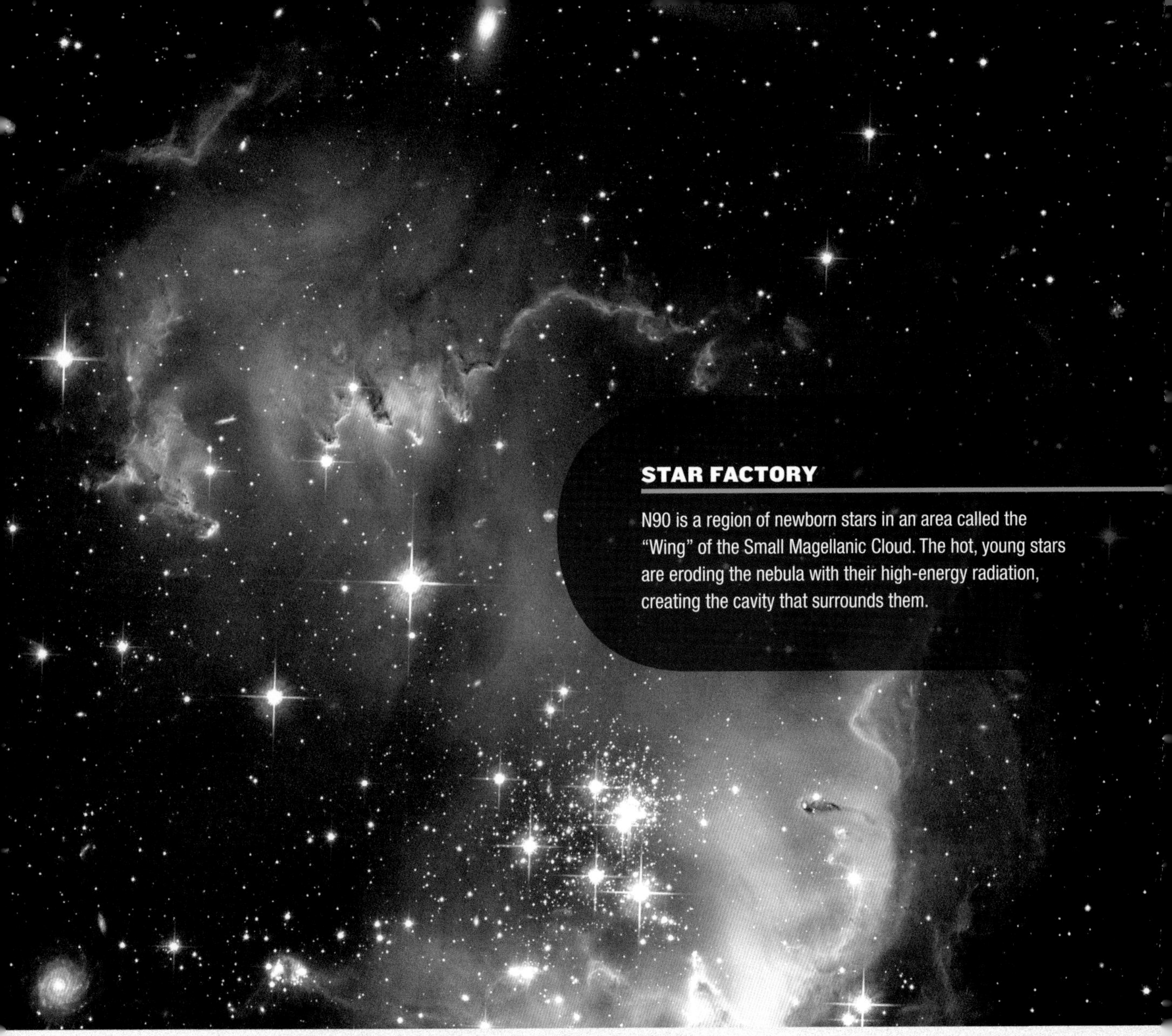

STAR FACTORY

N90 is a region of newborn stars in an area called the "Wing" of the Small Magellanic Cloud. The hot, young stars are eroding the nebula with their high-energy radiation, creating the cavity that surrounds them.

millennia that followed, increasingly high-powered telescopes and a deeper understanding of the cosmos have allowed us to identify different types of nebulae, and spot those that would be otherwise invisible by way of radiation and infrared detection methods. And thanks to space telescopes that orbit above the blurring effect of Earth's atmosphere, like Hubble, we can generate extraordinarily beautiful images of them.

As our views improved and widened, a new classification system was established based on the character of a nebula's light. Emission nebulae emit their own light, reflection nebulae reflect the light of hot stars nearby, and dark nebulae obscure glowing objects from our view. *Star Trek* explorers have further refined these distinctions, designating a multitude of classifications and types of nebulae. In our own universe, the 21st-century group of three stands.

SURPRISING SHAPES

Not all planetary nebulae are spherical, as the Bug Nebula, an envelope of gas ejected by a dying sunlike star, demonstrates. The cast-off material glows due to ultraviolet energy material from the star.

NEBULAE ABOUND

All nebulae have certain characteristics in common. They tend to be huge—sometimes hundreds of light-years across—and they're part of the interstellar medium, or the gases between stars. Nebulae can distinguish themselves in different ways: by their shape, their origin, how they're lit, and what elements they hold.

Emission (or glowing) nebulae are clouds illuminated by the heat of stars within. These newborn stars emit intense ultraviolet radiation that bathes the entire nebula, and the gases reemit the energy as light. Mutara class nebulae in *Star Trek* are intensely radioactive emission nebulae (VGR / "One").

While emission nebulae emit radiation and visible light, reflection nebulae shine by reflecting the light from surrounding stars. Dust scatters the nearby wavelengths at the blue end of the spectrum most easily, which often gives rise to an eerie glow. In the *Star Trek* universe, an emission nebula enshrouds the entire Amleth system and renders cloaking devices used inside its boundaries useless (DS9 / "Return to Grace").

Dark nebulae are so dense with gas and dust that they block visible or ultraviolet light. These clouds appear as irregularly shaped silhouettes set against a rich background of stars. Some dark nebulae—invisible to the human eye—appear bright in infrared, signifying a strong internal heat source that is likely from collapsing material that will form a star or stars. The result can create spectacular visual effects such as the famous, real-world Horsehead Nebula.

Radiation from hot stars illuminates and erodes this giant gaseous pillar known as the Cone Nebula, which resides in a turbulent star-forming region.

WHERE'S THE PLANET? WHERE'S THE STAR?

Despite their name, planetary nebulae don't actually have anything to do with planets. The astronomers who first observed them thought they looked like fuzzy, distant planets because of their spherical shape. They're actually a type of emission nebula that's left over when old red giant stars die and shed their outer layers, leaving a white dwarf core behind. Many of these are symmetric as early astronomers surmised, but advanced observations made with telescopes like Hubble have revealed that they often aren't spherical at all.

A massive supernova explosion leaves behind a supernova remnant. Their diffuse, quickly expanding gas sweeps up whatever material is in its path. They play an important role in interstellar infrastructure: Their magnetic turbulence is a major interstellar heating source; they're a source for many heavy elements like iron and gold; and they can trigger the next generation of star formation. A 10,000-year-old supernova remnant called the Cygnus Loop (also known as the Veil Nebula) may be on the older side, but its filaments are still expanding at about 60 miles (100 km) per second.

"As a matter of cosmic history, it has always been easier to destroy than to create."
—SPOCK

IN THIS MOVIE: Captain Kirk moves a mounting confrontation with his greatest enemy, the vengeful Khan Noonien Singh, to the Mutara Nebula, where both ships' sensors will be hampered by the nebula's effect on their shields. The Battle of the Mutara Nebula rages. Warp engines aboard the *U.S.S. Enterprise* reengage just in time to get it out of the nebula before the Genesis Device explodes, killing Khan. The nebula coalesces around the explosion site, creating the Genesis Planet. That is where they lay Spock to rest after he sacrifices himself to deadly levels of radiation in order to repair their warp drive.

AN INTERSTELLAR DUST CLOUD GLOWS INTO VIEW on the *Enterprise:* a vibrant blue and violently purple ribbon that curls and trembles in space, looking like a brilliant smear of paint against a sheet of glass. This is the Mutara Nebula, an emission nebula so distinct and intense in its activity that it inspired its own class of nebula in the *Star Trek* universe. Composed largely of ionized gases, it is constantly discharging high levels of electricity that flash like lightning bolts. The electrified gas renders electronic sensors unreliable and makes a starship's shields inoperable—effects that many a pilot uses to strategic advantage.

Emission nebulae make great intergalactic hiding places. The *U.S.S. Enterprise*-D avoids a Borg cube by camouflaging the ship inside a nebula (TNG / "The Best of Both Worlds"); Lieutenant Thomas Riker (posing as Will Riker) ducks the *Defiant* into an emission nebula while its cloaking devices aren't working (DS9 / "Defiant"); and Captain Ransom eludes a Klingon Bird-of-Prey for three days by hovering inside a nebula (VGR / "Equinox, Part II").

The Federation research lab Regula I is operated by Dr. Carol Marcus and orbits around D-class planetoid Regula near the Mutara Nebula.

WHALE OF A TALE

Many of Khan's most memorable quotes in Star Trek II: The Wrath of Khan *(1982) were paraphrases—or direct quotes—from Herman Melville's novel* Moby-Dick.

Expeditions throughout the *Star Trek* universe have brought to light more than ten types of nebulae, many more than the number classified in our universe. Mutara class nebulae tend to pose the greatest danger to passing ships. When the *Voyager* passes through a 110 light-year-long Mutara class nebula, a trip that will take them one month, most of the crew has to be put in stasis pods to protect them from exposure to skin-melting subnucleonic radiation (VGR / "One").

EMISSION NEBULAE IN OUR UNIVERSE

100 TO 10,000 TIMES THE MASS OF THE SUN / AVERAGE TEMPERATURE APPROXIMATELY 18,000°F (10,000°C)

When astronomers conduct studies of nebulae, they don't take just a single image. Telescopes like Hubble take many images, and then layer them together to produce a picture that is not only breathtaking, but can also tell us something about where various elements are located within the nebula.

EMISSION NEBULAE ARE COMPOSED OF THE SAME elements as the rest of the universe on the whole, including the stars, other types of nebulae, and the interstellar medium. Like most stars, they're primarily made of hydrogen, but emission nebulae also have a significant percentage of helium and a trace of heavier elements and other matter, including dust.

PINK WISPS

The hydrogen in an emission nebula is ionized, like *Star Trek*'s Mutara Nebula. One trait of ionized hydrogen is that it glows with a distinctly pink color at visible wavelengths, so any image of an emission nebula taken with a traditional camera's long exposure will be that color. But because astronomers are able to take images at all wavelengths, it would stand to reason that any well-equipped ship in the *Star Trek* universe would be similarly outfitted and would be able to analyze an emission nebula—or any celestial body—with ease.

INSIDE THE CLOUD

The density of an emission nebula will vary. Depending on a particular nebula's history, it could be millions of atoms per cubic inch or only a few atoms in the same space. The average density of a nebula is much less dense than any vacuum we can produce here on Earth, and far less than the density of Earth's atmosphere.

To light up like a neon sign with the characteristic glow that sets emission nebulae apart, the molecules within the cloud must be lit up by radiation from a very hot star. For a nebula composed mostly of hydrogen, this lighting occurs with a star that has a temperature of at least 44,540°F (24,727°C). Stars cooler than this don't have enough ultraviolet radiation to ionize or excite hydrogen atoms in the cloud to make them glow. The Orion Nebula (M42) is one of the most famous emission nebulae in our universe. Stars nestled deep in its

ARE WE THERE YET?

SHIELDS UP

A threatened starship automatically raises its shields, but real-life deflector shield technology is problematic. In 2014, students at the University of Leicester posited that plasma could be used to deflect some matter, but the ship inside wouldn't be able to see past it. Other researchers have come up with a plasma energetic field that, although it might block radiation, is probably no good against photon torpedoes.

Light-Years Away

Star Trek Generations

More than 3,000 stars appear in this image of the Orion Nebula, all of varying sizes. The bright region in the nebula's center is home to its four heaviest stars, collectively called the Trapezium.

heart flood the nebula's gases with UV radiation, turning it into a constant night light in our skies.

Though the average temperature of an emission nebula is around 12,000°F (6,649°C), its low density means that if you were to take a shuttlecraft through it, it would not melt your ship.

NAME RECOGNITION

Gene Roddenberry included "Noonien" in Khan's full name in honor of a friend he'd met in World War II, hoping the gesture would reunite them.

WEAPONS

Though the primary mission of many of *Star Trek*'s ships is exploration and research, the ships and crews are equipped with weapons in case they become necessary. Many of these weapons use beams of directed energy to stun, kill, or create an explosion, though less-advanced weapons like knives and swords are not uncommon.

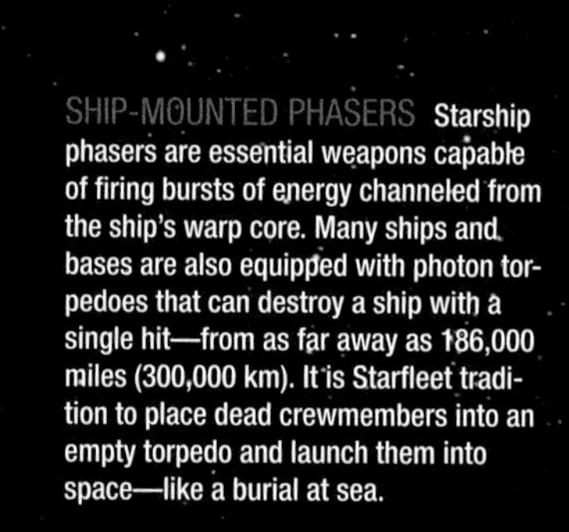

SHIP-MOUNTED PHASERS **Starship phasers are essential weapons capable of firing bursts of energy channeled from the ship's warp core. Many ships and bases are also equipped with photon torpedoes that can destroy a ship with a single hit—from as far away as 186,000 miles (300,000 km). It is Starfleet tradition to place dead crewmembers into an empty torpedo and launch them into space—like a burial at sea.**

HAND PHASER **The smallest and most discreet of Starfleet energy weapons, type 1 hand phasers are used in defense and as a cutting tool, explosive device, or energy source.**

PHASER PISTOL **Type 2 phasers are easily recognized by their distinctively pistol-like appearance, with a barrel and trigger not seen on the hand phaser.**

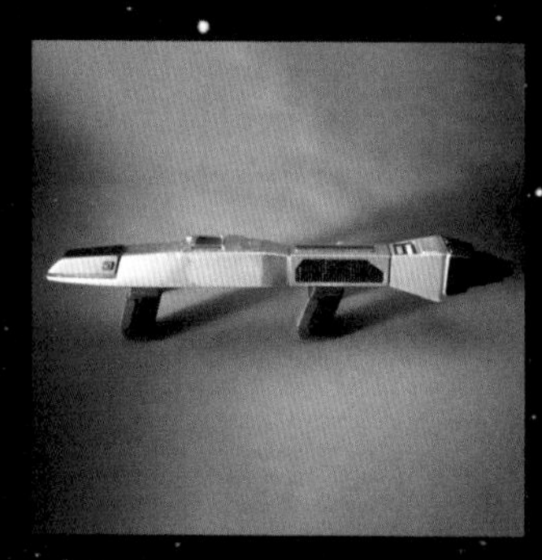

PHASER RIFLE **A more powerful weapon, the type 3 phaser rifle has 16 power settings ranging from a low-energy stun to high-energy phaser bolts.**

KLINGON DISRUPTOR The standard handheld weapon used by Klingons, this disruptor pistol has a distinctive design that matches other traditional Klingon weapons.

KLINGON *BAT'LETH* A traditional Klingon blade, the *bat'leth,* or sword of honor, is a popular weapon among Klingon warriors. According to Klingon mythology, the first bat'leth was forged by the legendary hero Kahless the Unforgettable. These crescent-shaped swords are often passed down from generation to generation.

KLINGON *D'K TAHG* Commonly used in hand-to-hand combat, the *d'k tahg* is the traditional knife of Klingon warriors. The hilt of this important ceremonial weapon often features a family crest.

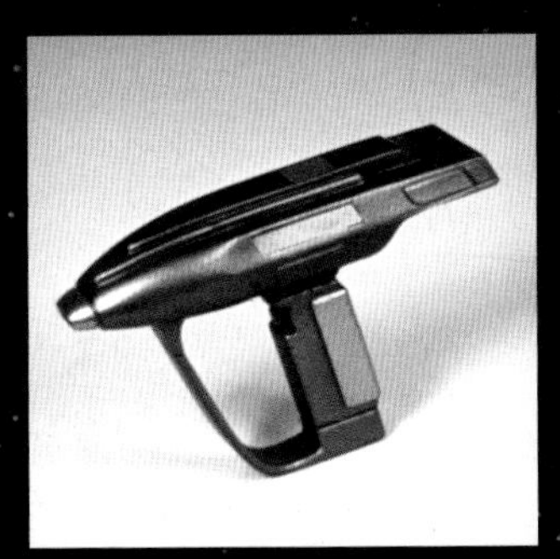

ROMULAN DISRUPTOR PISTOL Similar to Starfleet's phasers, disruptors are energy weapons. The Romulan disruptor pistol is a more powerful sidearm, with no stun setting.

CARDASSIAN DISRUPTOR PISTOL The simplest disruptor pistol, the Cardassian weapon of choice is sturdy and could be "dragged through the mud and still fire," according to Starfleet Commander Kira Nerys.

NAUSICAAN KNIFE Early in his Starfleet career, Jean-Luc Picard is stabbed in the heart with this Nausicaan knife. He survives the incident, but receives an artificial heart transplant.

EMISSION NEBULA IN THE NIGHT SKY

Look for the Great Orion Nebula within the constellation Orion.

STARGAZING TIPS

BEST VIEWING SPOTS: Visible in the Northern Hemisphere during winter months and the Southern Hemisphere during summer, but figure is upside down

BEST TIME TO SEE IT: Evenings, November to March

BASIC TIPS: The Great Orion Nebula, also known as Messier 42, is one of the most famous and easiest to find deep-sky targets for backyard stargazers. At only 1,340 light-years away, it is the closest large star-forming nebula to Earth. It shines at a magnitude +4.0, so it is easily glimpsed with the naked eye as a distinct small fuzzy patch of light, even from city suburbs. Binoculars will begin to reveal this central star in Orion's sword as a nebular mist. Views through telescopes, however, show the glowing stellar nursery in all its glory, looking much like a downturned flower in bloom.

HOW TO FIND IT

1. During early winter evenings look in the southern sky for three very noticeable medium-bright stars aligned in a perfectly straight row, representing the belt of the hunter Orion. From end to end these three stars stretch about 3 degrees across—equal to about the width of your two middle fingers held at arm's length.

2. The rest of the constellation is easy to trace: Above and below this striking stellar trio are even brighter stars that represent the shoulders and knees of the mythical nimrod. Orion's left shoulder is marked by the 500-light-year-distant red giant star Betelgeuse, while his right leg is pinned down by the 860-light-year-distant blue-white star Rigel.

3. Dangling below Orion's belt is a line of fainter stars just visible to the naked eye that form a hanging sword and lead you to the Orion Nebula. Two dim stars flank what looks like a fuzzy spot gleaming in between. This special "gleam" in the sword is the nebula, a giant swirling mass of gas. This star factory spans 24 light-years across space, and about 1 degree across Earth's sky—equal to the width of two full moon disks side by side.

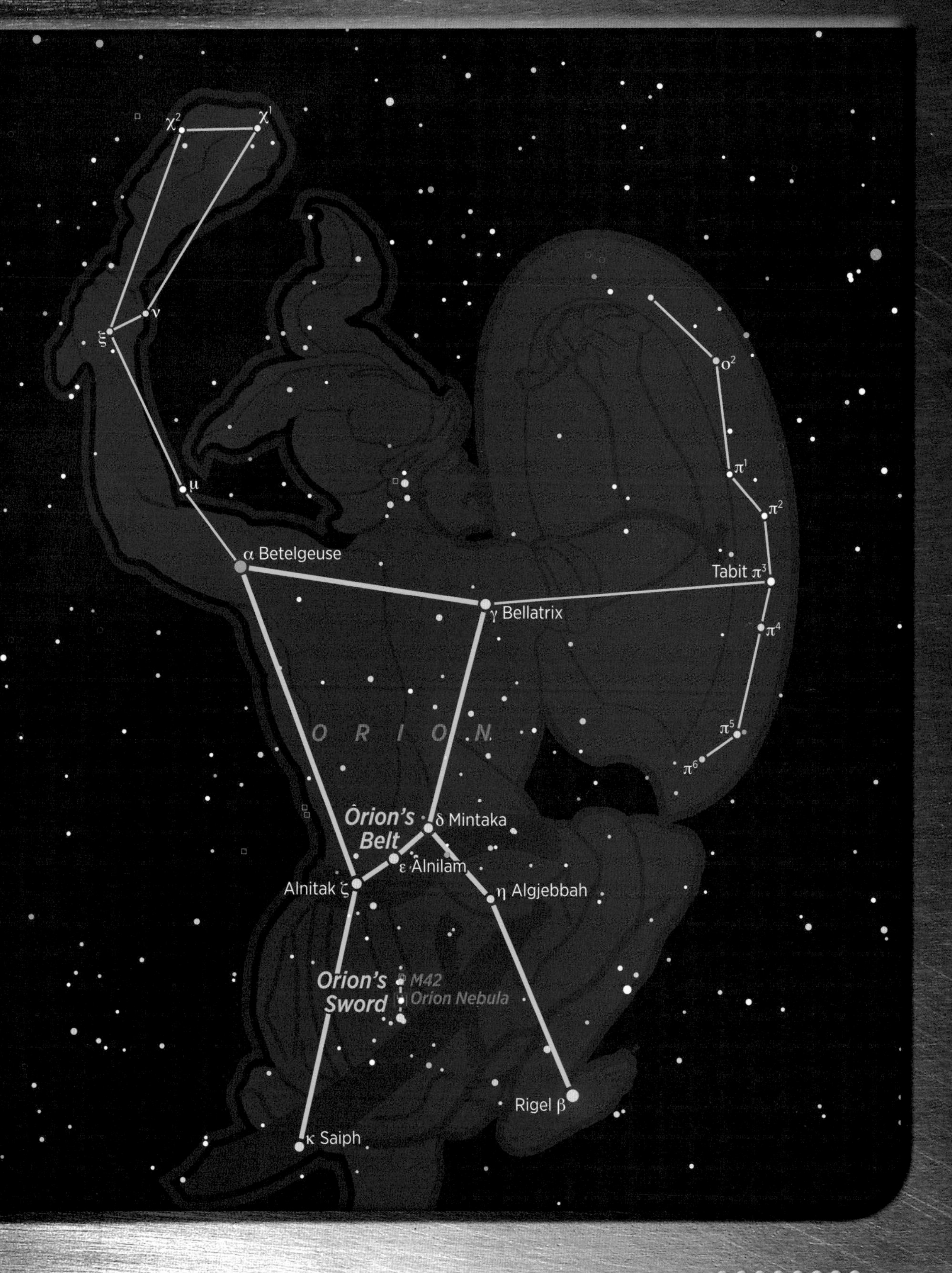
χ2
χ1
ν
ξ
μ
α Betelgeuse
γ Bellatrix
o2
π1
π2
Tabit π3
π4
π5
π6
O R I O N
Orion's Belt
δ Mintaka
ε Alnilam
Alnitak ζ
η Algjebbah
Orion's Sword
M42
Orion Nebula
Rigel β
κ Saiph

"I'd be delighted to offer any advice I can on understanding women. When I have some, I'll let you know."

—JEAN-LUC PICARD

DARK AND REFLECTION NEBULAE IN *STAR TREK*

TNG / "IN THEORY"
SEASON 4 / 1991

IN THIS EPISODE: Keen to learn more about the workings of Humans, Data pursues a romantic relationship with lonely crewmate Jenna D'Sora. While he tries to become her perfect partner, the *U.S.S. Enterprise*-D passes through Mar Oscura: a dark nebula that triggers odd behavior on the ship. The nebula is denser than others they've seen, filled with dark matter that Data postulates may have allowed strange life-forms to grow there. The nebula's deformations make it very difficult for the *Enterprise*-D to navigate, and Captain Picard has to use a shuttlecraft to guide them to safety.

AS THE ENTERPRISE-*D CRUISES TOWARD THE MAR* Oscura Nebula, it looms large on their viewscreens—huge wisps of black and blue swirl and churn like a boiling storm. Its giant, dense-looking tendrils of gas appear just as ominous. Dark nebulae like Mar Oscura are gaseous accumulations, and are difficult to detect with conventional sensors. That's why the *Enterprise*-D uses some modified photon torpedoes to light up the nebula, making it easier to navigate through. Even with these explosive lanterns, the nebula poses both mysteries and dangers. Mar Oscura's dense collections of dark matter create gaps in the fabric of space. The *Enterprise*-D triggers these gaps each time the ship accidentally runs into a dark matter patch, causing all sorts of anomalies. These anomalies—strange rearrangements and disappearances—seem manageable at first, until a crewmember is killed when she actually melts into the ship's deck.

Starfleet first sees a dark nebula directly in 2153 when Captain Archer and Subcommander T'Pol use metreon particles to light one up. They name it the Robinson Nebula in honor of Archer's old captain and friend (ENT / "First Flight"). Like other nebulae, their tendency to scramble sensors and thwart detection by enemy craft make dark matter nebulae tempting hiding places. Captain Sisko orders his damaged ship into a dark nebula in Cardassian space in 2374, but their inability to see clearly inside it means they soon find themselves crashed on an unknown planet (DS9 / "Rocks and Shoals").

The Enterprise*-D passes through ghostly dark nebula Mar Oscura, where high concentrations of dark matter distort objects on board in strange and dangerous ways.*

NEBULOUS BEINGS

When they enter a dark nebula in 2371, the* U.S.S. Voyager *stumbles upon noncorporeal beings called the Komar (VGR / "Cathexis").

C92

OOT. 1004.S27

DARK AND REFLECTION NEBULAE IN OUR UNIVERSE

DARK NEBULAE UP TO 150 LIGHT-YEARS ACROSS / INTERNAL TEMPERATURE OF A DARK NEBULA APPROXIMATELY -418°F (-250°C)

Dark and reflection nebulae don't emit their own radiation; even so, they are like emission nebulae in that they are named for how radiation affects them. Dark nebulae act as dust- and gas-laden curtains concealing stars and galaxies, whereas reflection nebulae are like mirrors shining light not their own.

DUST-DENSE DARK NEBULAE LEND A BIT OF mystery to the night sky, obscuring objects lying just beyond them and thwarting astronomers' attempts at full-spectrum studies of them. That isn't to say that dark nebulae are completely opaque. In fact, nebulae—even dark ones—are fairly transparent in the infrared segment of the spectrum. They are less so in the visible spectrum, which means you can look up and find the Southern Hemisphere's Coalsack Nebula silhouetted against a bright sky.

IN THE CLOUD

A dark nebula's composition is different than the rest of the low-density interstellar medium. Its hydrogen is in molecular form: H_2, the simplest possible molecule. The farther you travel toward its center, the colder and darker it becomes. Its density ranges from about 100 to 300 molecules per cubic centimeter, with a temperature from minus 440°F (-262°C) to minus 370°F (-223°C). Such density and cooler temperatures allow the hydrogen molecules to form there, making it an ideal place for stars to form after the nebula collapses.

The dust in the denser central regions of the cloud is great at blocking light. There is as much as 1/1,000 less starlight inside the cloud as outside it, making it an effective light filter. Though a dark nebula is likely too thin to effectively obscure a starship at close range—particularly if that ship had no way to conceal its infrared emissions.

THE HAZY MIRROR

Reflection nebulae are similar in composition to dark nebulae, with un-ionized gases intermingled with a haze of dust particles. But rather than blocking starlight from our view, reflection nebulae act as a hazy mirror reflecting starlight. They shine their borrowed light in our direction, much like water vapor reflects the light from our headlights back at us on a foggy night. But instead of water vapor, the ice-covered dust grains that are so efficient at blocking light in a dark nebula act as

ARE WE THERE YET?

INTELLIGENT MACHINES

When it comes to artificial intelligence, Lieutenant Commander Data is the ultimate dream machine. When it comes to putting this kind of cybernetics into a humanlike body, we've got a way to go. Still, artificial intelligence is making amazing strides: Robots can bluff, read human emotions, zip up your jacket, and chase after you, while a computer program can learn your behaviors and teach itself concepts.

Getting There

Android Data

countless tiny reflectors. Any dust grain with a bit of ferrous metal will align perpendicular to the magnetic field in the cloud.

Like the sky on Earth, reflection nebulae appear blue in the visible spectrum: The composition of the nebula scatters the short-wavelength blue light from a star more readily than the long-wavelength red light, making the nebula look blue. 

Reflection nebulae—like the Pillars of Creation—are so named because they reflect light from nearby host stars instead of glowing with their own light. The pillars are part of the Eagle Nebula in the constellation Serpens.

MOLECULAR CLOUD

LDN 988 is part of the Great Rift along the galactic plane, also called the Northern Coalsack. Stars are forming and igniting amid the clouds of dust, which are visible from dark sky sites on Earth. The nebula is located in the constellation Cygnus.

DARK NEBULA IN THE NIGHT SKY

Look for the Great Rift, a dark dust cloud that runs through the constellations Ophiuchus and Serpens.

STARGAZING TIPS

BEST VIEWING SPOTS: Dark countryside in the southern half of the Northern Hemisphere during the summer months, and the Southern Hemisphere in winter

BEST TIME TO SEE IT: Late evenings, May to October

BASIC TIPS: On dark summer nights away from the light pollution of cities, a long dark gap appears to fracture the ghostly glowing band of the Milky Way. Known as the Great Rift, this dark cosmic lane begins in the southern constellation Sagittarius, the Archer, and runs up the sky through Ophiuchus, the Serpent Bearer, to around the border of Cygnus, located nearly overhead in mid-northern latitudes.

HOW TO FIND IT

1. Look toward the southern sky for a bright orange star 550 light-years distant—Antares, the lead member in the constellation Scorpius, the Arachnid.

2. To the left of Antares is the great Teapot asterism or star pattern in the Sagittarius constellation, near the southern horizon. Between them lies the soft band of hazy light that is the Milky Way and contains the beginning of the Great Rift, which continues to stretch toward the overhead sky.

3. Scan above Antares for the 2.4 magnitude star Sabik in the faint constellation Ophiuchus. The two stars are separated by about 14 degrees—slightly less than the span between your little finger and index finger held at arm's length. Follow Sabik and a line of faint 3rd and 4th magnitude stars from Nehushtan to Alya, off to its upper left. Continue hopping the line of stars that leads straight into the heart of the dark gap, called the Great Rift, which splits the Milky Way band down the middle.

Rasalhague α
κ
β Cebalrai
Tsung Ching γ
Alya θ
λ Marfik
η Donghai
O P H I U C H U S
δ Yed Prior
Yed Posterior
ε
Great Rift
ν Yan
υ She Low
Han ζ
ν
Serpens
ξ Nehushtan
η Sabik
φ
χ
ψ
ρ
θ
Antares
Teapot
Milky Way

"If I didn't know better
I'd say this ship is trying
to kill me."
—CHAKOTAY

IN THIS EPISODE: The *U.S.S. Voyager* shuts down power as they near an eerie-looking nebula, casting the ship's halls in shadow. To keep them entertained, Neelix tells some former Borg children the story of why Deck 12 is closed off. He tells them how, while collecting deuterium in a similar nebula months before, they accidentally picked up a ghostly hitchhiker that turned the ship against them—isolating them in darkness and flooding them with poisonous gas. Captain Janeway had two choices: Reason with the alien presence or consign her ship to its destruction.

THE NEBULAE THAT THE VOYAGER *HEADS TOWARD* in this episode, both past and present, shimmer like an apparition—glimmering in translucent wisps of pink, purple, and blue. Both nebulae are categorized as class J, rich in the isotope deuterium (hydrogen with a proton and a neutron) that's crucial to fueling all of the ship's matter-antimatter reactions. These interstellar gas clouds offer the crew a vital resource, but they still find them unsettling to travel through. That may be because nebulae in *Star Trek* are notoriously mysterious, known for wreaking havoc with a ship's systems and for containing unexpected objects and unknown species.

Categorizing nebulae in *Star Trek* can be a tricky business. They're often labeled based on the effect they have and the materials they contain: fictive sirillium and real-life argon, helium, and deuterium, the last of which can be found in the nebulae of our universe. Starships make collecting deuterium look easy with the use of their Bussard collector, but mining for it in our universe would be akin to trying to pan for gold in a river.

When the Voyager *mines a nebula for deuterium, the crew collects more than they bargained for: an alien presence that means them harm.*

A NEBULA'S NET

The Delta Flyer *ejects its damaged warp core into a class J nebula, knowing its ionized gases will contain the warp's explosion (VGR / "Drive").*

When it comes to planetary nebulae—remnants left over from a big star's death with that star's dead core left at its center—*Star Trek* doesn't specify which of its nebulae fall under this title. Ships like the *Enterprise* NX-01 are known to survey such nebulae (ENT / "Breaking the Ice"), and planetary nebulae such as the Ring Nebula are often displayed on a starship's large screens in full, rich color (VGR / "Alice").

PLANETARY NEBULAE IN OUR UNIVERSE

1,500 DETECTED IN MILKY WAY, AN ESTIMATED 3,000 EXIST
LIFE SPAN OF A FEW THOUSAND YEARS

Our sun will become a red giant when it reaches the end of its life, eventually collapsing and expelling its outer layers in a shell known as a planetary nebula. This glowing envelope will last a few tens of thousands of years before dissipating back into the interstellar haze.

UNLIKE MANY TYPES OF NEBULAE IN THE STAR TREK universe, planetary nebulae are not categorized by their content or their effect on visitors from afar. Rather, they are classified by their origin and appearance. When Charles Messier recorded this type of nebula—27th on his famous list in 1764—he described it as oval and said that it contained no star. His contemporary William Herschel called it "planetary" simply because it appeared so round and about the same size as a planet in the sky, but Herschel's son John discovered that planetary nebulae have structure. In 1833, he renamed Messier 27 the Dumbbell Nebula.

Only about one-fifth of nebulae are rounded, and most aren't symmetric. They come in a variety of shapes that resemble things like hourglasses, butterflies—even clusters of shimmering bubbles. Recent images from Hubble have revealed an array of amazingly complex structures with a central white dwarf, all that's left of a star that was once similar to our own sun. There are an estimated 3,000 of these briefly glowing wonders in the Milky Way, none of which are visible to the naked eye, but rather hidden in the interstellar dust.

Although every star except our sun appears as a point of light in a telescope, planetary nebulae have a sizable disk with a measurable diameter. The disk is brighter toward the edges; this ringlike illusion is due to the fact that we can see a greater concentration of gas at the edges of a three-dimensional nebula than through the middle.

Beautiful planetary nebulae are ideal objects to display, as seen on the *U.S.S. Enterprise*'s bridge monitor (TOS / "Charlie X"). But they also hold an interesting blueprint to a star's evolution and chemical composition over the course of its life.

A SPHERE OF ATMOSPHERE

These planetary nebulae are born out of the dying process of stars containing less than eight solar masses, meaning that their life cycle after the main

ARE WE THERE YET?

SOMETHING FROM NOTHING

Want to grab yourself a cup of Picard's preferred beverage? Go to the starship's replicator, say "Tea, Earl Grey, hot," and a cup will appear. Today's 3-D printers may not be able to make you a good cup of tea, but they're getting there. They can create guns, building materials, and human cells. They can even produce chocolate and crackers that will sprout a salad.

Mission Accomplished

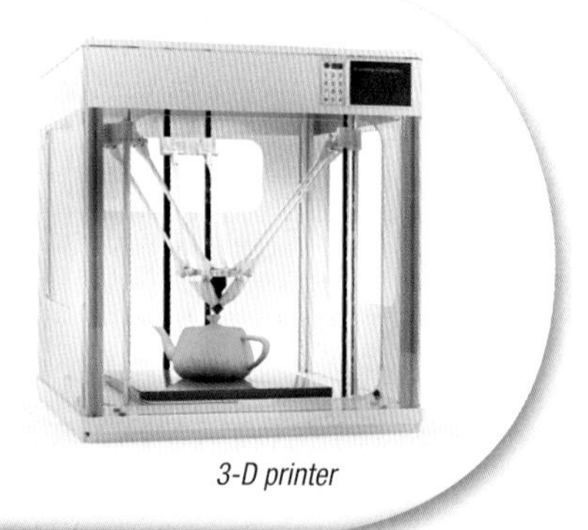

3-D printer

The Dumbbell Nebula's delicate halo is seen from Earth in the constellation Vulpecula, thanks to clouds of hydrogen and oxygen radiating around a white dwarf star. Our own sun may have a similar fate some 5 billion years from now.

sequence will lead them to evolve into red giants. After the star spends a small percentage of its lifetime in this bloated red giant state—about 10 to 20 percent of its active life—explosions in its helium-burning shell will produce energy pulses. These forces push the outer envelope of the stellar atmosphere outward, away from the star within. The hot naked core of the dying star bathes the gases in the ejected envelope with incredible amounts of hot, ultraviolet light. This ultraviolet radiation lights the planetary nebula from within like an enormous fluorescent lightbulb.

Like all nebulae, they are subject to outside forces acting to move their gases around and disperse them into the interstellar medium. As such, planetary nebulae retain their distinct shapes for only a few tens of thousands of years. The stellar core at the nebula's center will cool during this time, and a white dwarf will mark the place where the nebula used to be.

HOLOGRAPHIC TECH

In the *Star Trek* universe, holograms are a crucial form of technology for communication, investigation, and even entertainment. Within the confines of a holodeck (seen in background) or a projector, specialized force fields hold together what is known as holomatter, a substance that, under computer control, can mimic the properties of real matter. Some holographic programs, such as the Doctor, are even considered sentient beings.

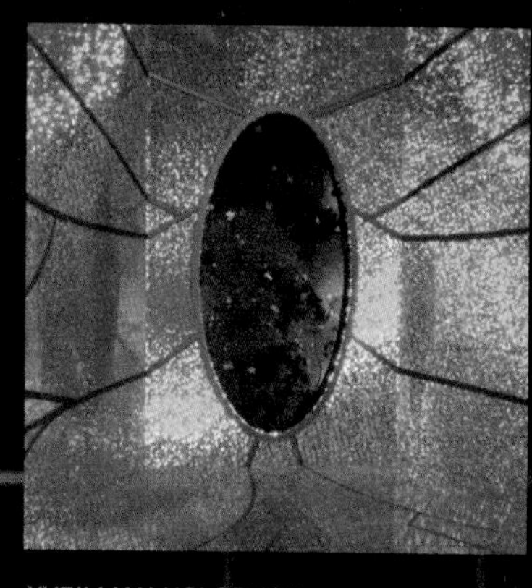

XYRILLIAN HOLODECK **An encounter with a Xyrillian vessel introduces Starfleet to advanced holographic technology. Their holographic chamber inspires Starfleet's creation of its own holodeck.**

HOLOFILTER (DS9) **A device used to hide a person's identity, the holofilter masks the speaker's physical appearance during communications. Here, Commander Benjamin Sisko is seen through the holofilter as a Kobheerian to bypass Cardassian patrols.**

HOLO-COMMUNICATOR (DS9) **Holograms bring communication off viewscreens and into the room. Using projectors in the floor, the holo-communicator can send and receive holograms to allow two-party conversations.**

THE DOCTOR Though intended only for extreme cases, *U.S.S. Voyager*'s Emergency Medical Holographic program, the Doctor, serves as full-time chief medical officer during its unexpected voyage through the Delta Quadrant.

PLANETARY NEBULA IN THE NIGHT SKY

Look for the Dumbbell Nebula, within the constellation Vulpecula.

STARGAZING TIPS

BEST VIEWING SPOTS: The Northern Hemisphere in summer, and low in the sky in the Southern Hemisphere during winter months

BEST TIME TO SEE IT: Late evenings, May to October

BASIC TIPS: The Dumbbell, also known as Messier 27, is considered the finest planetary nebula visible to backyard sky watchers. Shining at a very faint 7.4 magnitude, it is best found and observed under dark skies, and with binoculars it looks like a tiny oblong-shaped glow with a distinct brighter central region. Under high magnification using telescopes, the nebula takes on more of a classic hourglass or apple core shape filled with intricate filamentary details.

HOW TO FIND IT

1. Look for the giant stellar pattern of the Summer Triangle near the overhead sky. At about the halfway point between the triangle point stars Vega and Altair, on the inner side, is the very faint naked-eye +4.4 magnitude star Anser, the lead member in the tiny constellation Vulpecula, the Little Fox.

2. From Anser hop to +4.6 magnitude 13 Vulpeculae, which is just over 5 degrees away—a separation equal to the width of three middle fingers held at arm's length. Pick up the Dumbbell Nebula using binoculars when sweeping 2 degrees south of 13 Vulpeculae—equal to a thumb's width.

3. As an alternative star hop, begin the hunt in the slightly brighter small constellation Sagitta, the arrow that lies between the constellations Cygnus and Aquila south of Vulpecula. Star hop to the Dumbbell sitting only 3 degrees north of +3.5 magnitude Gamma Sgr.

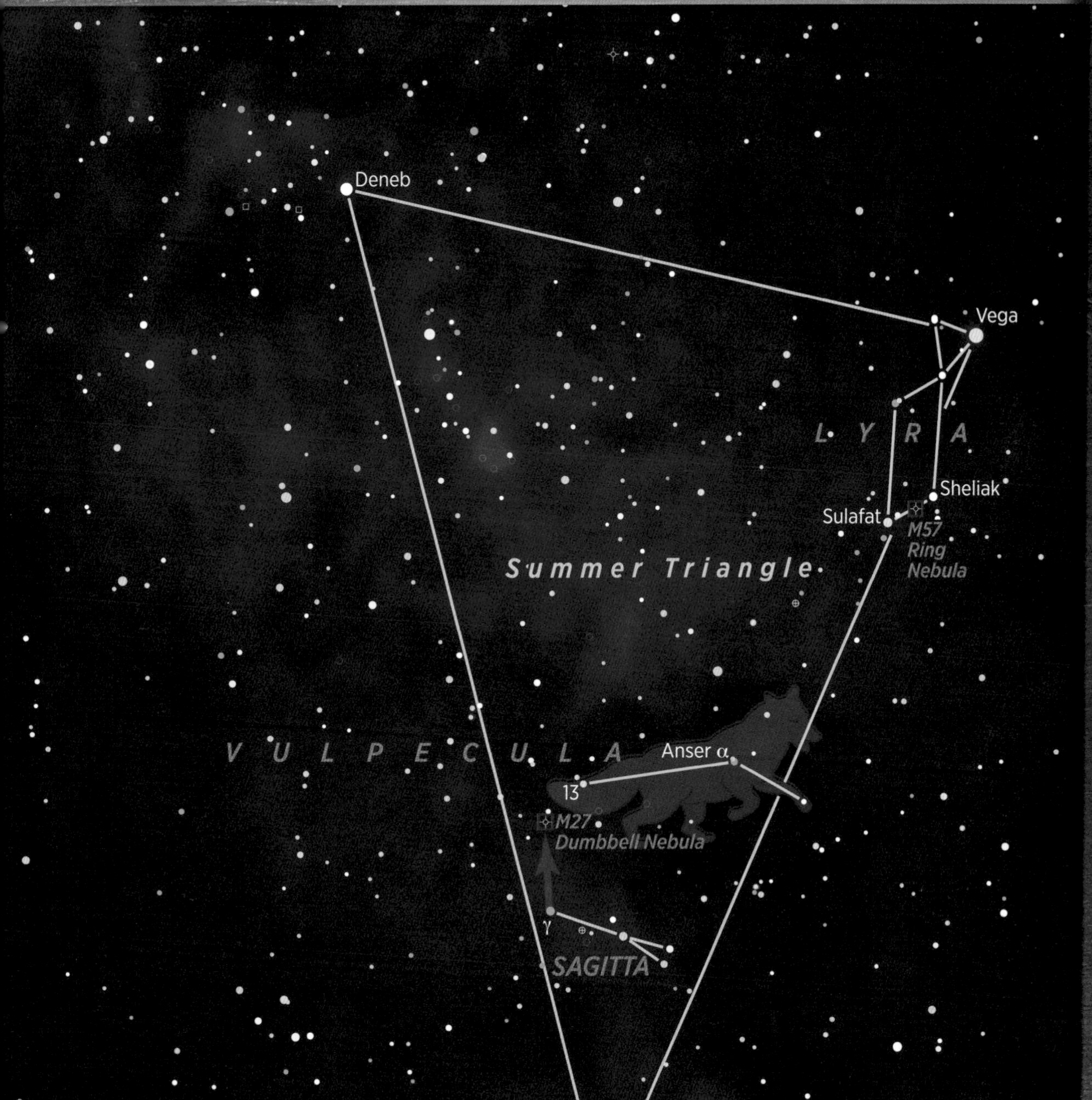

Deneb
Vega
LYRA
Sheliak
Sulafat
M57
Ring
Nebula
Summer Triangle
VULPECULA
Anser α
13
M27
Dumbbell Nebula
γ
SAGITTA
Altair

"If the Enterprise *hadn't jumped into warp when it did, we would have been blown to pieces."*

—GEORDI LA FORGE

SUPERNOVA REMNANTS IN *STAR TREK*

TNG / "EMERGENCE"
SEASON 7 / 1994

IN THIS EPISODE: When two holodeck programs merge together in strange ways, the crew of the *U.S.S. Enterprise*-D uncovers a strange phenomenon: The ship has begun to reproduce. This process turns out to be a difficult one, and the ship uses the holodeck program to try to communicate to the crew what help it needs. In response, the *Enterprise* heads toward the MacPherson Nebula, where they detonate a modified photon torpedo to mine it for vertion particles. These particles are what the ship requires to feed its circuit nodes; once its nodes are fed, the ship is able to finish creating its offspring.

FROM AFAR, THE MACPHERSON NEBULA LOOKS like hazy purple storm clouds with a faint white glow at its center—almost like the smoke of a long-ago fire left to mark the site of a spectacular event. Indeed, it marks the spot where a supernova once exploded through space, leaving this ghostly remnant behind.

La Forge guides them to the MacPherson Nebula; he has to choose between it and a pulsar, deciding which one he thinks is more likely to yield the materials the ship needs to finish reproducing. Though the materials found inside nebulae sometimes prove hazardous, they can also prove extremely useful. When they fire their modified torpedo into this nebula, it shimmers like a jewel as they extract its vertion particles.

Supernova remnants appear to hover like ghosts in the *Star Trek* universe, offering up clues to events long past and making them prime candidates for further exploration. In both *Star Trek* and real life, when a supernova explodes it leaves behind ejected material bounded by an expanding shock wave that engulfs and shakes whatever materials it touches. *Star Trek*'s supernova remnants sometimes overlap with those in our own universe: The Veil Nebula, captured by our Hubble telescope, also pops up on the astrometrics display on the *U.S.S. Voyager* (VGR / "Year of Hell"). In *Star Trek*'s negative universe, a supernova event actually wakes up rather than kills a dead star (TAS / "The Counter-Clock Incident")—something dead stars in our own universe just cannot do.

The Enterprise*-D crew modifies one of their photon torpedoes so they can use it to mine a nebula for vertion, which the new life-form needs to survive outside the ship.*

SEEKING A REMNANT

Captain Janeway uses her desire to study a supernova remnant as a convenient reason to enter an off-limits section of Devore space (VGR / "Counterpoint").

EXPANDS FOR UP TO SEVERAL HUNDRED YEARS AFTER SUPERNOVA
LIFE SPAN OF ABOUT 10,000 YEARS

A supernova event is violent yet soundless; its aftereffect is just as silent and just as spectacular. Ghostly supernova remnants are ejected material bound by the supernova's shock wave that expands at almost 19,000 miles a second (30,578 km/s), shocking whatever material it happens to swallow.

THERE ARE A FEW MAIN TYPES OF SUPERNOVAE, and the remnant of each has its own characteristics. A Type Ia supernova is the result of a specific set of circumstances. This supernova type occurs during a zombie star's second death—that is, when a white dwarf in a binary system gravitationally pulls enough mass from its companion that it reaches 1.4 times the mass of our sun, known as the Chandrasekhar limit. At this mass, a white dwarf collapses on itself. This reheats it as internal pressures rise, igniting a new round of fusion and causing a supernova, and then a nebula. Little to no hydrogen will be present in this nebula: The vast majority of its mass will belong to the former white dwarf, which had barely any hydrogen to begin with.

Another hydrogen-deficient remnant is the aftermath of a Type Ib supernova, which is what is left after a star has lost its hydrogen shell and exploded. But the star death that most often comes to mind when someone says "supernova" is the Type II supernova. That's what occurs when a single massive star has completed its dying supergiant phase and detonates. Unlike either of the Type I supernovae, a Type II supernova remnant will have hydrogen in it, because the star was able to hang onto its hydrogen envelope.

EVOLUTION

After the moment of death, the supernova remnant—in the case of Type Ib and Type II, at least—will fade and cool significantly after a few years. The shell of gas expands and cools, and as the heat dissipates, the material condenses into grains of dust. Radiation and stellar winds will push the remnant around, and over tens of thousands of years it will expand into the interstellar medium, mingling with clumps of gas and dust in other clouds.

ASHES TO ASHES

Supernova ashes play an important role in the formation of new generations of stars and planetary systems.

ARE WE THERE YET?

PHASER TECH

***Star Trek*'s phaser weapons can do things like dematerialize, disrupt, and stun. Today, real directed energy systems transmit scorching heat to a target by way of a one-kilowatt solid-state beam. Boeing's Laser Avenger can knock drones from the sky or detonate explosives from a distance. A more powerful truck-mounted system is also in development to combat threats from rockets to mortars.**

Getting There

Laser technology

The majestic Veil Nebula is a supernova remnant from a star that exploded between 5,000 and 10,000 years ago. Energy from the explosion continues pushing outward and creating shock fronts that heat the gas to millions of degrees.

A remnant's expansion can push into a dark nebula, setting off compression and collapse. Over time, the grains will stick together, forming progressively larger and denser clumps, becoming dense knots of gas. These protostars can be concealed by nebulae that will eventually form part of the star's mass. Eventually—if the conditions are right—the protostar may gain enough mass to ignite and begin fusion, providing light and the potential for life.

MINING NEBULAE

The Delta Flyer II *turns to a nebula as a potential source of dilithium, a crystalline mineral that helps control and power warp drive (VGR / "Nightingale").*

CRAB NEBULA

The star that perished in a supernova in 1054 C.E., which created the Crab Nebula, became a pulsar. The colors represent different elements in the nebula: Blue is neutral oxygen, green is singly ionized sulfur, and red is doubly ionized oxygen. The nebula spans about ten light-years.

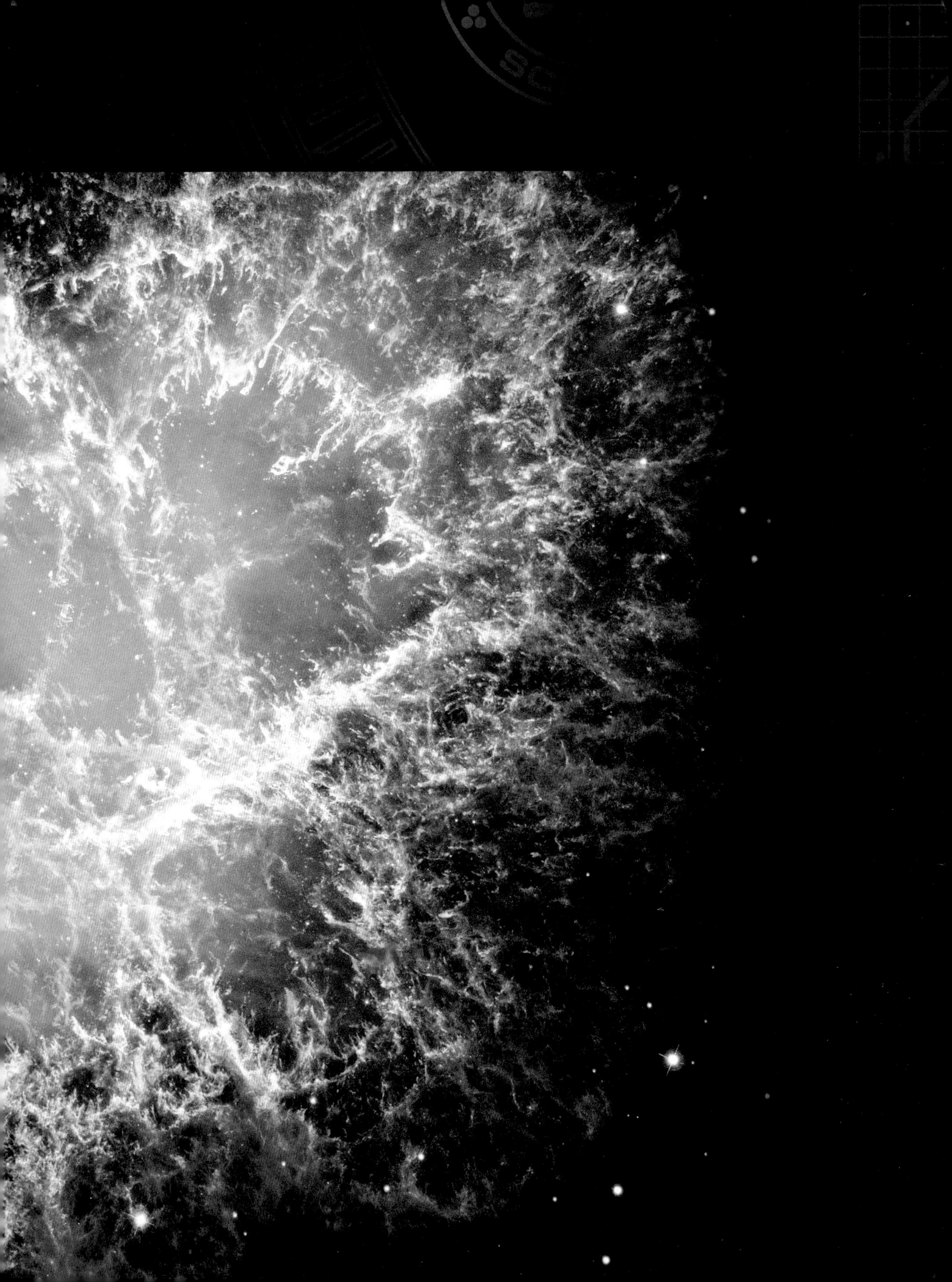

SUPERNOVA REMNANT IN THE NIGHT SKY

Look for the Crab Nebula within the constellation Taurus.

STARGAZING TIPS

BEST VIEWING SPOTS: Northern Hemisphere in autumn and winter, and late spring and summer in the Southern Hemisphere

BEST TIME TO SEE IT: Evenings, October to April

BASIC TIPS: The Crab Nebula, also known as Messier 1, is the brightest supernova remnant visible to stargazers and sits about 6,500 light-years from Earth. It is found in Taurus, the Bull, one of the largest and most prominent constellations best visible in winter across the entire Northern Hemisphere. Situated near the tip of the southern horn of the mythical beast, the Crab shines at magnitude +8.4. It can be spotted with strong binoculars as a tiny fuzzy glow, but using a telescope's magnification power will begin to reveal it as an oval cloud filled with intricate details.

HOW TO FIND IT

1. Start by drawing an imaginary line to the right of the three bright stars of Orion's belt until you reach the next brightest star, the orange-colored Aldebaran. This 65-light-year-distant red giant is the brightest member of the constellation Taurus. Aldebaran marks the left tip of a V-shaped pattern of stars.

2. Continue an imaginary line from the base of the V-shaped cluster, through Aldebaran and to the next brightest star called Zeta Tauri or Tien Kwan, a naked-eye +3.0 magnitude star that marks the tip of the bull's southern horn. The irregular oval-shaped fuzzy path of light that is the Crab Nebula is located 1.1 degrees northwest of Tien Kwan, or equal to the width of the little finger held at arm's length.

3. As an alternative way to locate this supernova remnant, draw an imaginary arc between superbright orange star Betelgeuse in Orion to Tien Kwan and Elnath, the bright naked-eye +1.7 magnitude star that marks the tip of Taurus's other horn. Again, from 440 light-year Tien Kwan, swing about 1 degree off this arc to find the magnificent Crab.

Elnath β

M1
Crab Nebula

Tien Kwan ζ

Pleiades

Ain ε

Hyades Cluster

δ[1] Secunda Hyadum

Aldebaran α

θ

γ Prima Hyadum

λ

TAURUS

ξ

ο

ν

Betelgeuse

Orion

Orion's Belt

CLUSTERS AND GALAXIES

BEYOND THE MILKY WAY ARE COUNTLESS GALAXIES TO EXPLORE, IF ONLY WE COULD CROSS THE VAST DISTANCES BETWEEN US AND THEM.

CHAPTER FIVE

STAR TREK & US

Billions of stars have been charted across *Star Trek*'s Milky Way galaxy, with billions more yet to be discovered. Warp technology could enable starships to do the unthinkable—to leave the confines of the galactic barrier, the mysterious negative energy shield that surrounds the Milky Way, and journey to exotic places that provide new insights into the universe.

THE STAR-STUDDED BEYOND

Yet even with warp speed, starships have only charted a fraction of the Milky Way; the Gamma and Delta Quadrants remain largely unexplored. The immense distances that separate the Milky Way from the nearest large galaxies, measured in millions of light-years, take starships moving at maximum speeds hundreds of years to traverse.

In our universe, we've mapped less than 300 million stars. That's just a fraction of *Star Trek*'s billions, but a valiant effort given the cosmos's gargantuan size. Astronomers began the largest stellar mapping effort to date in the early 21st century, using the 8.2-foot (2.5-m) mirrored Isaac Newton Telescope in the Canary Islands to plot more than 200 million stars across the northern Milky Way, some a million times fainter than what the human eye can see. In 2014, the Gaia mission began surveying stars and other astronomical objects from space, providing better insight into the distances and motions of stars in the Milky Way galaxy than ground telescopes ever could.

We've learned through ever widening searches that many of these stars huddle together in enormous clusters. In *Star Trek*, these clusters contain exciting possibilities. The *U.S.S. Enterprise*-D visits one such cluster that's home to the Rubicun system. The cluster hosts more than 3,000 habitable planets, including Rubicun III, home to the friendly Edo. Captain Picard's crew is the first to make contact with the Edo in 2364 (TNG / "Justice").

A BUSTLING DISK OF LIGHT

In both *Star Trek* and our universe, the Milky Way is distinguished by a disk of stars, marked by spiral arms that create a sort of pinwheel formation. This dusty disk contains thousands of stars, many of them gravitationally bound into open star clusters or globular clusters. Open star clusters harbor anywhere from a few dozen stars to impressive collections of thousands strewn across the black tapestry of 10 to 20 light-years. The compact, spherical balls known as globular clusters hold populations tallying in the hundreds of thousands to millions. These rarer cosmic baubles swarm in a halo above and below the galactic plane.

PREVIOUS PAGES: The Deep Space 9 station, originally known as Terok Nor, lives near the mouth of the Bajoran wormhole and provides Starfleet with a defensive stronghold and a commercial port.

STELLAR HUDDLE

What appears to be a single cluster is more likely two clusters merging. This composite cluster is found in the Tarantula Nebula in the Large Magellanic Cloud. The shapes have become elongated, resembling what happens when two galaxies merge.

HOW CLUSTERS FORM

Open clusters form as stars coalesce out of giant molecular clouds of gas and dust. The stars are thus roughly the same age, but vary in mass as well as the number and conditions of their orbiting planets, if there are any. In *Star Trek*, Starfleet sometimes finds life amid these clusters. The *U.S.S. Enterprise*-D discovers life on an orbiting planet in the Zeta Gelis star cluster it surveys in 2366. The alien—an amnesiac Zalkonian whom they name John Doe—is all the more surprising in a star cluster they once believed to be uninhabited (TNG / "Transfigurations").

Life is possible in any real-life stellar system if the conditions are right, and it helps to be among older, smaller stars that are likely to live long and die quietly, avoiding the flood of radiation when large stars go supernova.

TYPES OF GALAXIES

Galaxies come in different shapes and sizes. Our home galaxy, the Milky Way, is a barred spiral. The next closest full-size galaxy to us is Andromeda, a spiral galaxy. Beyond that are at least 100 billion other galaxies, each a different size.

COSMIC SENIORS

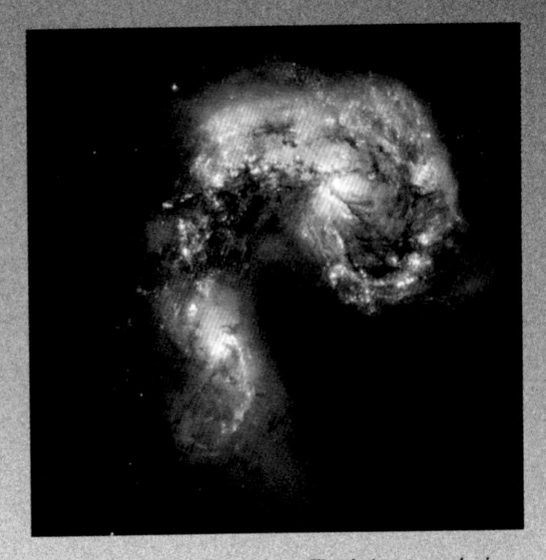

The Antennae galaxies are a pair of colliding galaxies. The merge will spur an era of dramatically increased star production.

Although open clusters are found in the disk of a galaxy, globular clusters are mostly confined to its halo. Globular clusters (or globulars) tend to be spherical and many times larger than their open cluster counterparts, stretching up to 100 light-years across. Made up of as many as a million or more stars, they can be so dense that identifying individual members is difficult even through a powerful telescope. In reality, their stars are separated by about a light-year—a globular cluster's compactness is an illusion created by distance. Explorers in the *Star Trek* universe find them difficult to navigate thanks to gravimetric interference, but that doesn't stop them from trying.

Globulars are thought to be cosmic old-timers that formed about 12 billion years ago, probably as part of the process that gave rise to the Milky Way itself.

ISLANDS IN SPACE

How a galaxy's stars and clouds are distributed determines its form. The Milky Way, for example, is a 100,000 light-year-wide galaxy with spiral arms streaming from its center. It has enough dense clouds of dust and gas to make billions more stars to add to its existing 200 to 400 billion. Our sun's solar system lies buried within one of the galaxy's outer arms, about 30,000 light-years from the center.

Astronomers classify galaxies by three basic shapes: elliptical, disk (many with spirals), or irregular. The Milky Way is a barred spiral galaxy that is true to its name: a spiral galaxy, with a central region that's a dense bar of stars.

In *Star Trek*, some lucky starships get close-up glimpses of real-life galaxies: Triangulum, our next closest spiral galaxy after Andromeda, calculated to take approximately 300 years to reach at maximum warp; inhabited Andromeda; and NGC 2812, nestled in the Cancer constellation and visible from Earth.

GALACTIC METROPOLIS

Like stars, galaxies will also flock together to form clusters. Our Milky Way is part of what is known as the Local Group, a cluster that contains at least 54 known galaxies of varying sizes spread across seven million light-years. Typically, the smaller irregular galaxies orbit the larger spirals like the Milky Way and Andromeda galaxies. The memberships of these galactic clubs range from less than a dozen to titanic assemblages of thousands of galaxies stretching across hundreds of millions of light-years.

These galactic distances are so vast that even the *Star Trek* universe's explorers have made little more than a ripple in the cosmos. Yet future generations of *Star Trek* explorers—and, maybe someday, our universe's own—will continue to push technology's boundaries, creating machines in which we can boldly go where no one has gone before.

“Concentrate. Make the disk go into the cone.”

“How do I do that?”

“Just let go.”

—ETANA JOL AND RIKER

OPEN STAR CLUSTERS IN *STAR TREK*

TNG / "THE GAME"
SEASON 5 / 1991

IN THIS EPISODE: When Wesley Crusher arrives for a visit aboard the *U.S.S. Enterprise*-D, he finds the crew behaving strangely. They've all become addicted to a Ktarian game that Wesley suspects is bending their minds in more ways than they know. The game makes the crew vulnerable to a plot by Ktarian Etana Jol, who planted the game on Will Riker in hopes of using mind control to take over the *Enterprise*-D. As they head toward the uncharted Phoenix cluster, Wesley and his new friend Robin Lefler try to save the crew from the game's control while trying not to fall victim to it themselves.

FROM AFAR, THE PHOENIX CLUSTER FEATURES A bright white core and a burst of light surrounded by a smattering of pale, small-looking stars. It looks quaint from the *U.S.S. Enterprise*-D's vantage point, but the crew remarks that its size is actually quite impressive. They take on extra engineers for the task of exploring it, and Riker comments that the two weeks they've been given won't be nearly enough time to do the cluster justice. This trip in 2368 makes the *Enterprise*-D the first Federation starship to explore the Phoenix cluster, and the crew is excited to do so.

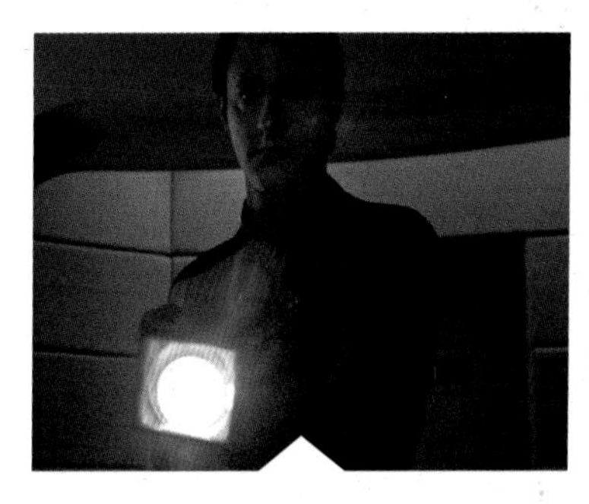

BEACON OF LIGHT

Data alters a palm beacon to break the crew out of the Ktarian game's dangerous and addictive trance, but it's usually used as a handheld flashlight.

Star clusters—groups of stars held close together by gravitational forces—are a favorite destination for Starfleet explorers. The fact that the stars in these clusters are born around the same time means they are great places to study stellar evolution. They sometimes include habitable planets, so they can also be a great place to establish Human colonies. In 2364, the Federation establishes one such colony inside a 3,000-star stellar cluster in the Strnad system (TNG / "Justice"). Another massive cluster is the Dorias cluster, which includes over 20 star systems, several of which have habitable planets within them (TNG / "Bloodlines").

They may be ripe for exploration, but according to *Star Trek*, clusters like these can be so dense that getting ships through them takes a practiced hand. Clusters like the Argolis cluster are dangerous to pass through because of the gravimetric shear often associated with them (DS9 / "Behind the Lines").

Taking a break from Starfleet Academy, Wesley Crusher is transported from an Oberth*-class starship to the* Enterprise*-D for a visit with the crew.*

OPEN STAR CLUSTERS IN OUR UNIVERSE

100 TO 1,000 MEMBER STARS
DIAMETER OF UP TO 30 LIGHT-YEARS
ABOUT 1,000 IN MILKY WAY

In the real universe, as in the *Star Trek* universe, star clusters are perfect laboratories for examining the evolutionary paths of stars. Starfleet's crews get to conduct these studies up close, while Earth's astronomers must observe them from afar.

RATHER THAN TAKING READINGS FROM THE SHIP as it travels through a system, we can look at real-life star clusters as a group and learn a lot about them. Stars in open clusters have similar ages, born within a few million years of each other. Because they're born from the same cloud of dust, the chemical composition of these stellar siblings tends to be about the same. Open clusters' difference arises in their masses, which can vary depending on the original nebula's variations, as well as whether or not they go on to become part of multiple star systems.

A LOOSE-KNIT FAMILY

Young open clusters live within the galactic disk and range from 5 to 75 light-years across. Because they contain between a dozen and several thousand stars, they are more loosely packed than the other class of bigger star clusters, called globular clusters. This impression of comparative "openness" is enhanced because the globular clusters are much farther away from us and out of the galactic plane. From Earth, we can see large, nearby open clusters like the Pleiades and the Hyades with the naked eye.

All the stars in an open cluster will dance in an orbit around a common center of gravity: the center of the cluster's mass. This orbiting happens somewhat slowly compared to the speed of a starship traveling through it, so it wouldn't make it more difficult to navigate.

GAUGING AGE

To determine the approximate age of a cluster, astronomers can examine each star and plot its position on the Hertzsprung-Russell diagram. As the member stars evolve, they will begin to peel away from the main sequence, beginning with the hottest and most massive and continuing down and to the right along the diagram's main sequence line (see p. 107), through the cooler, less massive stars. By looking

ARE WE THERE YET?

VIRTUAL REALITY

Stepping into the holodeck is like stepping into another world, turning fantasy into almost-reality. We may not be able to enjoy an experience where matter is impersonated, but virtual experiences are almost upon us. Oculus Rift will make users feel like they've entered a new reality. Other companies are developing programs that project images beyond your TV and onto your walls.

Getting There

Samsung Gear virtual reality goggles

Pleiades is the easiest open star cluster to spot with the naked eye. These young, bright blue stars are siblings in a true sense, as they were born from the same molecular cloud.

at what point in evolution the shift to the next stage begins—which is called the "main sequence turnoff point"—we can tell the age of the rest of the stars. For older clusters, the turnoff point is going to be fainter and redder than younger clusters. This method makes it much easier to determine the age of stars that are members of a cluster than to gauge the age of a star that is all by itself.

BUSTLING CLUSTERS

Star clusters are often crowded places: The Dorias cluster contains both a dichromatic nebula and a class 4 pulsar (TNG / "Bloodlines").

THE FIRST WARP-CAPABLE STARSHIP In the *Star Trek* universe, the *Enterprise* (XCV-330) is the earliest vessel to be given starship status. It is an integral part of Earth's burgeoning space travel program. It is the only 21st-century ship to reach its destination, Alpha Centauri, a neighboring star of the Sol system.

OPEN STAR CLUSTER IN THE NIGHT SKY

Look for the Big Dipper, some stars of which belonged to a former open cluster that has since come apart, within the constellation Ursa Major.

STARGAZING TIPS

BEST VIEWING SPOTS: Throughout the Northern Hemisphere, including temperate latitudes where it is located in the northern sky, and partially seen from northern parts of the Southern Hemisphere

BEST TIME TO SEE IT: Evenings, March to September, but can be visible year-round

BASIC TIPS: The seven brightest stars in the large northern constellation Ursa Major, the Great Bear, form probably the best known pattern of stars or asterism visible to the unaided eyes. Known as the Big Dipper or the Plow, it is composed of three stars that make the handle, and four that are the bowl. Its pattern also traces the bear's hindquarters and tail while the remainder of the constellation, which is the third largest, is composed of fainter stars that are more challenging to glimpse from bright suburbs. The Big Dipper's two end stars in the bowl are used as pointers.

HOW TO FIND IT

1. The Big Dipper can be found high in the northwestern evening sky in spring and low in the northeastern evening sky in autumn. For observers north of latitude 40° N (New York, Madrid, Beijing), Ursa Major is circumpolar, meaning it is always above the horizon and visible every night of the year.

2. Look for three stars that form what looks to many like a soup ladle's handle (Alkaid, Mizar, and Alioth), and the four stars that are the bowl (Megrez, Phecda, Merak, and Dubhe).

3. Use the Big Dipper stars to guide you to other neighboring bright stars and their constellations. Try taking the Big Dipper's two end stars in the bowl, Merak and Dubhe (called the Pointer Stars), and extend a line north five times the distance between the two stars until you reach the next brightest star, Polaris, or North Star, that marks the end of the Little Dipper, a similar, much smaller asterism within the tiny constellation Ursa Minor, the Little Bear.

Polaris
(North Star)
Little Dipper

Big Dipper
ο Muscida
α Dubhe
υ
Mizar ζ
ε Alioth
Megrez δ
η Alkaid
β Merak
Phecda γ
Talitha
Borealis
Al Haud θ
ι
URSA MAJOR
κ
Talitha
Australis
χ Alkafzah
ψ Ta Tsun
λ Tania Borealis
Tania Australis μ
Alula Borealis ν
Alula Australis ξ

"The cluster's a lot more dense than we thought. It's going to take three days just to map one-tenth of it."

—GEORDI LA FORGE

GLOBULAR CLUSTERS IN *STAR TREK*

TNG / "SCHISMS"
SEASON 6 / 1992

IN THIS EPISODE: As the *U.S.S. Enterprise*-D sets out to study the globular cluster known as the Amargosa Diaspora, Will Riker goes to sleep and wakes up exhausted. Then a subspace pocket forms in the cargo bay and crewmembers start going missing. La Forge and Data find a rift within the subspace pocket with something on the other side—something extending so deep into subspace that it shouldn't exist in their universe. Riker has to figure out how to escape the aliens experimenting on him and his crewmates and close the rupture before it's too late.

THE AMARGOSA DIASPORA—A DENSE, BULGING series of purplish clouds surrounding what looks like a globe of white light—looks like it would be a real hassle to navigate through. It's so massive that the *Enterprise*-D has to modify warp speed just to try to speed up the process of mapping it. It's actually this modification that allows solangen-based life-forms to create a subspace pocket on the ship and experiment on its crew. Their subspace rift shuts down technology like La Forge's Visual Instrument and Sensory Organ Replacement (VISOR) and Data's internal chronometer, making it more difficult to make it safely through the cluster.

A globular cluster is usually spherical in shape (its name comes from the Latin *globulus*, or small sphere), and has a lot more stars than an open cluster. It's held together tightly by gravity and is dense with ancient stars. Unlike our universe's dust- and gas-free globular clusters, the Amargosa Diaspora is surrounded by clouds. Ships navigating through them often experience gravimetric interferences, so Riker has to help Sariel Rager plot a course.

The Enterprise-*D heads toward an unusually dense globular cluster called Amargosa Diaspora, so star-packed and vast that trying to map it seems impossible.*

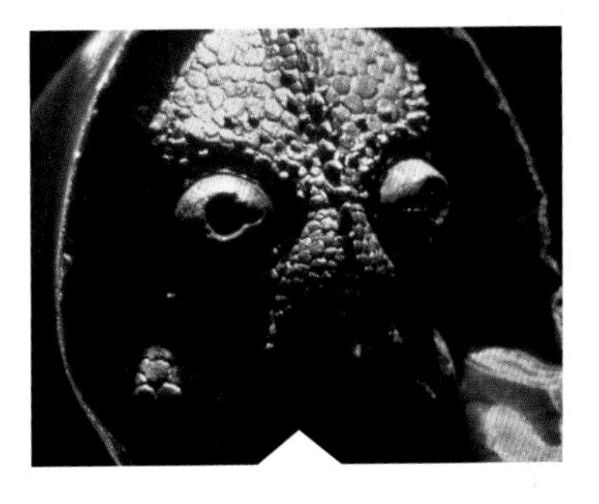

FIRST ABDUCTION

"Schisms" (TNG) marks the first incident in Star Trek *where Humans are actually abducted by aliens—at least in the most popular sense.*

Star Trek features tons of clusters, but it isn't always clear whether or not they're globular. Starfleet occasionally has to contend with clusters that don't even exist in our own universe. Black clusters, created when bunched protostars collapse, absorb energy in ways that wreck a starship. In 2368, the *Enterprise*-D is almost destroyed when it is sent to the edge of such a cluster (TNG / "Hero Worship").

GLOBULAR CLUSTERS IN OUR UNIVERSE

MORE THAN 150 IN MILKY WAY
FIRST RECORDED IN 1665 / ON AVERAGE, ABOUT 25,000 TIMES BRIGHTER THAN THE SUN

If an open cluster is like a family, a globular cluster can be compared to an isolated island with a very large, yet close-knit community. Everyone is related, and they interact much more with each other than with the outside world.

GLOBULAR CLUSTERS DIFFER FROM OPEN CLUSTERS in several respects. For starters, these balls of stars are more populous, possibly containing more than a million stars. Stars in tight clusters are also more likely to influence each other gravitationally. The intricate dance that occurs in globular clusters can get exciting; occasionally, a star will be pulled so forcefully that it reaches escape velocity and is ejected from the cluster and its system, left to wander the space between the stars as a halo field star.

INSIDE THE SPHERE

The high density in globular clusters means that the view from a planet orbiting one of these stars would be incredible. Imagine a night sky full of a million visible stars, compared to the 3,500 or so that we can see from a dark site on a moonless night. Sadly, the potential for stargazing from such a vantage point could be minimal. Low amounts of planet-building elements like iron and silicon and the crowded environment in globular clusters means that many researchers think planets would be rare in those systems—only one has been found as of 2016. But some researchers argue that because exoplanets have been found around stars that contain only one-tenth the metal content of our sun, planet formation in globular clusters might be possible. The long-lived, cool, red dwarfs that are left in these clusters also have closer-hugging habitable zones where orbiting planets would be relatively safe from stellar interactions.

THE MISSION CONTINUES

Studies of these clusters are ongoing. One target, IC 4499, was observed to determine whether its stars formed at about the same time or if it contained multiple generations of stars. It has long been held that globular clusters contain only old stars, with no new star formation within them. When an observation of IC 4499 conducted in the 1990s suggested that there were young stars present, astronomers were puzzled. Decades later, the Hubble Space Telescope has

ARE WE THERE YET?

HOLD STILL

Hyposprays make for convenient space medicine: They don't penetrate the skin, and they work through clothing to distribute medicine. Real-life "jet injectors" are already a part of our reality, though to date they haven't worked quite as well as their *Star Trek* equivalent. Newer jet injectors inject medicine at different depths and doses without needles, making them an increasingly enticing option.

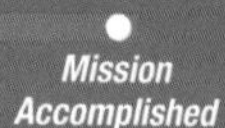
Mission Accomplished

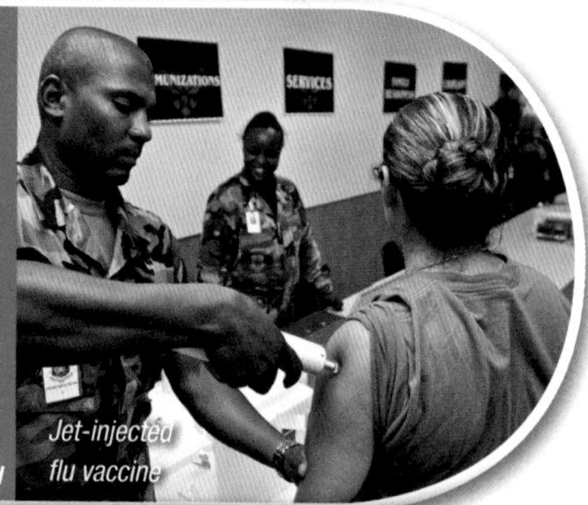

Jet-injected flu vaccine

provided far more precise estimates of star age, revealing that the stars in IC 4499 are indeed closer to the age of other clusters in the Milky Way, and supporting the idea that less massive clusters generally only consist of a single stellar generation. 

Hundreds of thousands of stars live within the Great Hercules globular cluster, also known as Messier 13 (M13). Its cluster type is spherical and dense, home to stars that are nearly as old as the universe itself and stretch across a 145 light-year diameter.

OMEGA CENTAURI

This stunning small region of Omega Centauri reveals the range of potential star colors and demonstrates the Hubble Space Telescope's Wide Field Camera 3's incredible color-capturing versatility. Omega Centauri is the largest globular cluster in the Milky Way.

GLOBULAR STAR CLUSTER IN THE NIGHT SKY

Look for the Great Hercules cluster, a globular cluster, within the constellation Hercules.

STARGAZING TIPS

BEST VIEWING SPOTS: Under darker skies in the Northern Hemisphere and Southern Hemisphere

BEST TIME TO SEE IT: Late evenings, April to October in the Northern Hemisphere, and in June to September, low in the north of the Southern Hemisphere

BASIC TIPS: The Great Hercules cluster, also known as Messier 13, is one of the finest examples of a globular visible from mid-northern latitudes. It is barely visible to the unaided eye under dark skies as a tiny +5.8 magnitude hazy spot buried within the large but faint constellation Hercules, the Strongman. Binoculars show it as a fuzzy ball, so this magnificent deep-sky treasure—lying an impressive 22,200 light-years away—is best observed through a telescope. Higher magnification can begin to resolve the individual stars along the cluster's edge and show off its brighter core.

HOW TO FIND IT

1. Facing the eastern sky in spring or the high south in the summer, look for the bright star Vega, the westernmost point of the Summer Triangle, and Arcturus, the bright orange star to which the Big Dipper's handle points.

2. Trace an imaginary line from Vega to Arcturus. At approximately a third of the line's distance are four moderately bright 3rd magnitude naked-eye stars that form a lop-sided rectangle pattern. This asterism, called the Keystone, marks the chest of Hercules.

3. Use binoculars to more precisely pinpoint Messier 13's location, about a third of the distance along the line from Eta Herculis to Zeta Herculis.

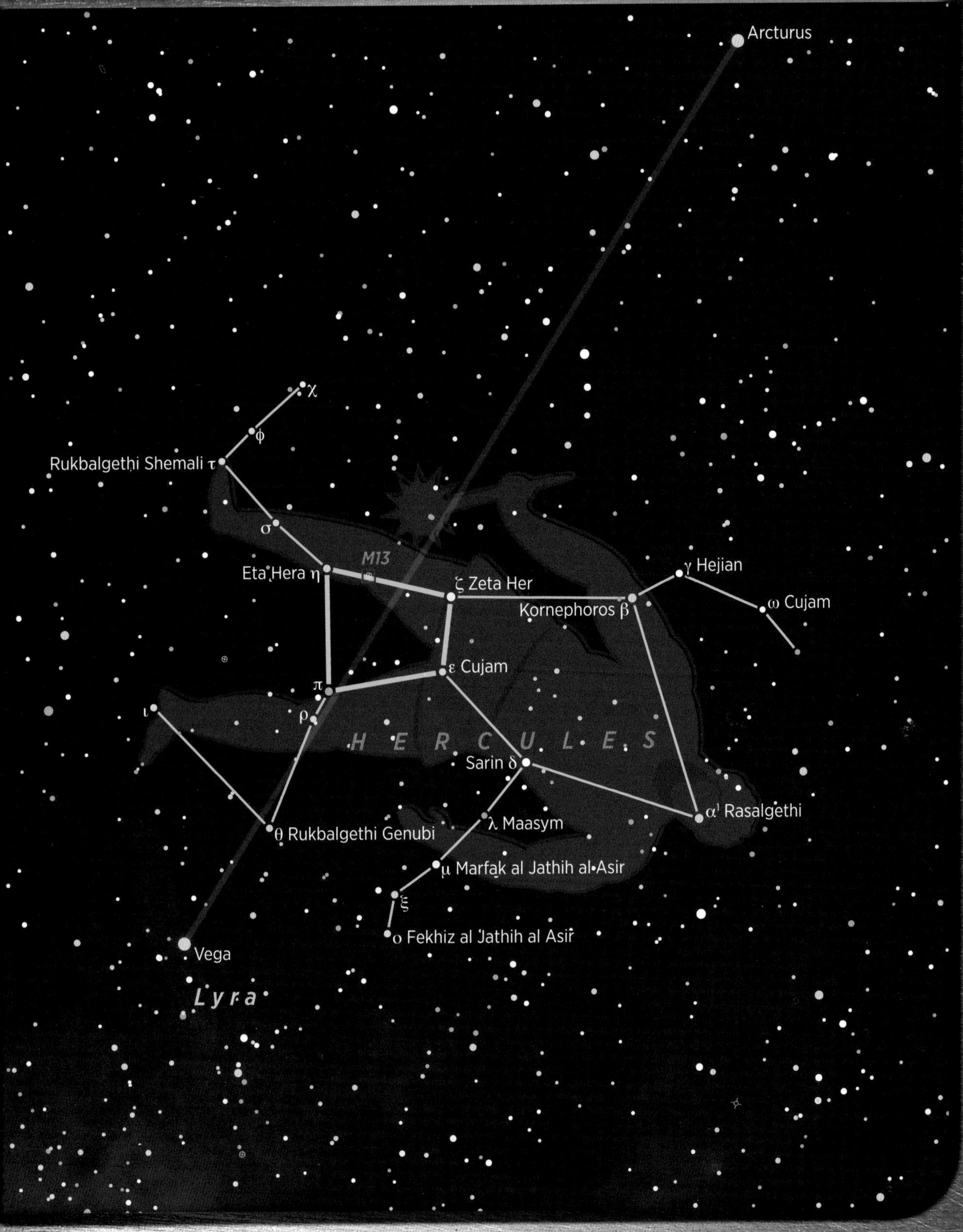

Arcturus

χ

ϕ

Rukbalgethi Shemali τ

σ

M13

Eta Hera η

ζ Zeta Her

γ Hejian

ω Cujam

Kornephoros β

ε Cujam

π

ι

ρ

H E R C U L E S

Sarin δ

α1 Rasalgethi

λ Maasym

θ Rukbalgethi Genubi

μ Marfak al Jathih al Asir

ξ

ο Fekhiz al Jathih al Asir

Vega

Lyra

"Yes, but where is this place?"

"Where none have gone before."

—JEAN-LUC PICARD AND DATA

IN THIS EPISODE: After an experimental modification to the ship's engine, the *U.S.S. Enterprise*-D is hurtled 2,700,000 light-years through space. The ship and its crew end up in the Triangulum galaxy, so far from home it would take them 300 years at maximum warp to get back. Captain Picard orders the engineer who enhanced their engines to send them back, but instead they end up traveling a billion light-years in the wrong direction—farther than anyone has gone before. Their situation is so dire that they have to rely on a mysterious alien called the Traveler to guide them home.

AS THE ENTERPRISE-D SLOWS AFTER A WARP acceleration that Data exclaims is "off the scale," the crew finds themselves in a mysterious place of dazzling color. Silk-like swaths of blue and purple hang among blue and bright purple star clusters. The ship sails past a ringed purple planet that looks much like Saturn; a bright white-purple orb hangs tantalizingly in the distance.

The crew's first stop on their superfast accidental journey, the Triangulum galaxy (or M33), is located about three million light-years from the Alpha Quadrant. It's beyond the Andromeda galaxy, far from Starfleet's home turf—but not quite as far as where they end up after their second bizarre warp experience, which puts them in a place where anomalies bloom on viewscreens and strange apparitions appear. Data argues that they should stay and study this new frontier containing a giant, still forming protostar, but the crew feels out of its depth.

Star Trek's home galaxy, like ours, is so huge and takes so long to travel through that starships rarely venture outside of it. The Milky Way's closest neighbor is our sister spiral, the Andromeda galaxy, home to two satellite galaxies and the Kelvan Empire. The Kelvans break through the galactic barrier surrounding the Milky Way because radiation levels threaten to make their galaxy uninhabitable. They make the trip in three centuries, a feat that amazes Captain Kirk because it would take a Federation starship thousands of years to reach Andromeda (TOS / "By Any Other Name").

After an engine experiment goes awry, the Enterprise*-D finds itself 300 light-years from home in the M33 galaxy, otherwise known as the Triangulum galaxy.*

TORTOISE AND THE HARE

The Enterprise*-D needs 300 years to travel 2.7 million light-years, whereas the* Voyager *needs 70 years to travel 70,000, making it nine times faster.*

Most of the action takes place within the Milky Way galaxy's Alpha Quadrant—except for *Voyager,* which ended up in the Delta Quadrant. To travel outside of the Alpha Quadrant takes dramatic transportation methods, because galaxies are big enough to hold billions of stars, and the space between galaxies is even larger.

GALAXIES ARE A VAST CONGLOMERATION OF objects: the interstellar medium of gas and dust, nebulae, asteroids, planets, stars, and stellar remnants, as well as a supermassive black hole at its center and a dark matter halo. The cosmos's massive, busy galaxies are recognized by their shape, and fit in three main shape-inspired categories: disk (sometimes called spiral), elliptical, and irregular galaxies. They can contain a few million to several billion stars, and can span a few thousand light-years across to more than a hundred times that. Exactly how many galaxies our universe holds is still unknown, but in 2012 the Hubble eXtreme Deep Field photo revealed a shocking number—approximately 5,500—in only one patch of sky. Estimates set the number of galaxies in the universe at a mind-blowing 100 to 200 billion.

GETTING IN SHAPE

Disk galaxies are a flat, disk-shaped arrangement of gas and dust with a glowing central bulge in the middle. Spiral arms can often be seen radiating from a central hub in the disk, which is why this type is also often known as a "spiral galaxy." A subtype of the spiral galaxy is the barred spiral, like the Milky Way, which features a bar rather than a ball of bright stars at its center. This type has many different colors of stars within, both old and young.

Elliptical galaxies are more spherical, and look much like the central bulge of a disk galaxy. They vary widely in size, from the rare, massive, giant ellipticals to the small dwarf ellipticals, which are the most common type of galaxy in the universe.

Irregular galaxies don't take a particular shape, so this is a common type for galaxies that may be the debris of a recent merger or collision of larger galaxies. This category includes the nearby Magellanic Clouds, which might have interacted with the disk of the Milky Way in the last 100 million years, and may—or may not—circle around to merge with the Milky Way in the distant future. Irregular

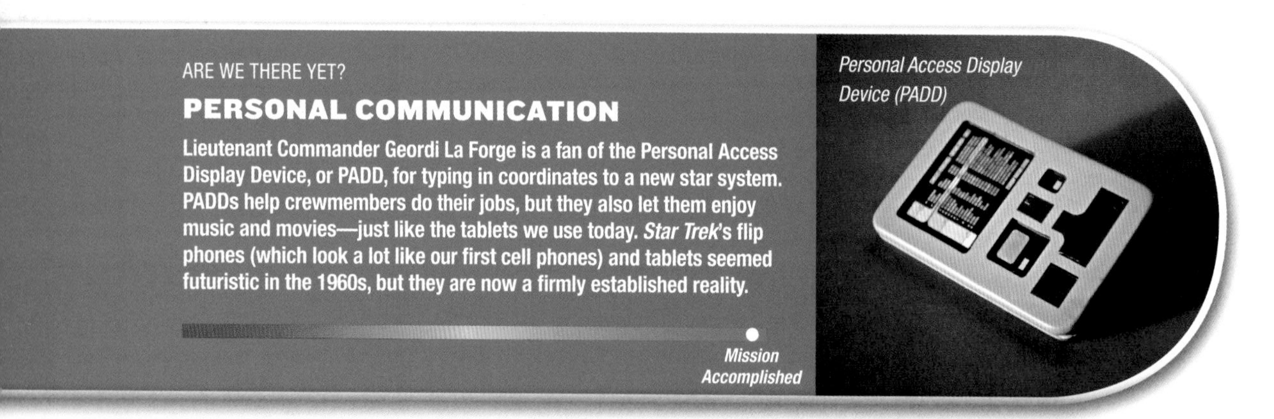
ARE WE THERE YET?

PERSONAL COMMUNICATION

Lieutenant Commander Geordi La Forge is a fan of the Personal Access Display Device, or PADD, for typing in coordinates to a new star system. PADDs help crewmembers do their jobs, but they also let them enjoy music and movies—just like the tablets we use today. *Star Trek*'s flip phones (which look a lot like our first cell phones) and tablets seemed futuristic in the 1960s, but they are now a firmly established reality.

Mission Accomplished

Personal Access Display Device (PADD)

galaxies often contain hot, young stars made from their plentiful clouds of gas and dust, and were more common when the universe was younger.

Two and a half million light-years distant, the Andromeda galaxy is seen at a steep angle from Earth. In 1912, it was the first object to be found to have a redshift that indicated it was not in our galaxy.

SOMETHING OLD, SOMETHING NEW

Over time, a galaxy may begin as one type, and as it merges with other galaxies, become another, transforming from a pair of disk galaxies into one (twice as large) elliptical galaxy. In addition to distorting the galaxies' shapes, these mergers can also throw off irregular galaxies, or spark a cascade of new star formation, as the gases and dust in the interstellar mediums of both galaxies smash together to create new stars.

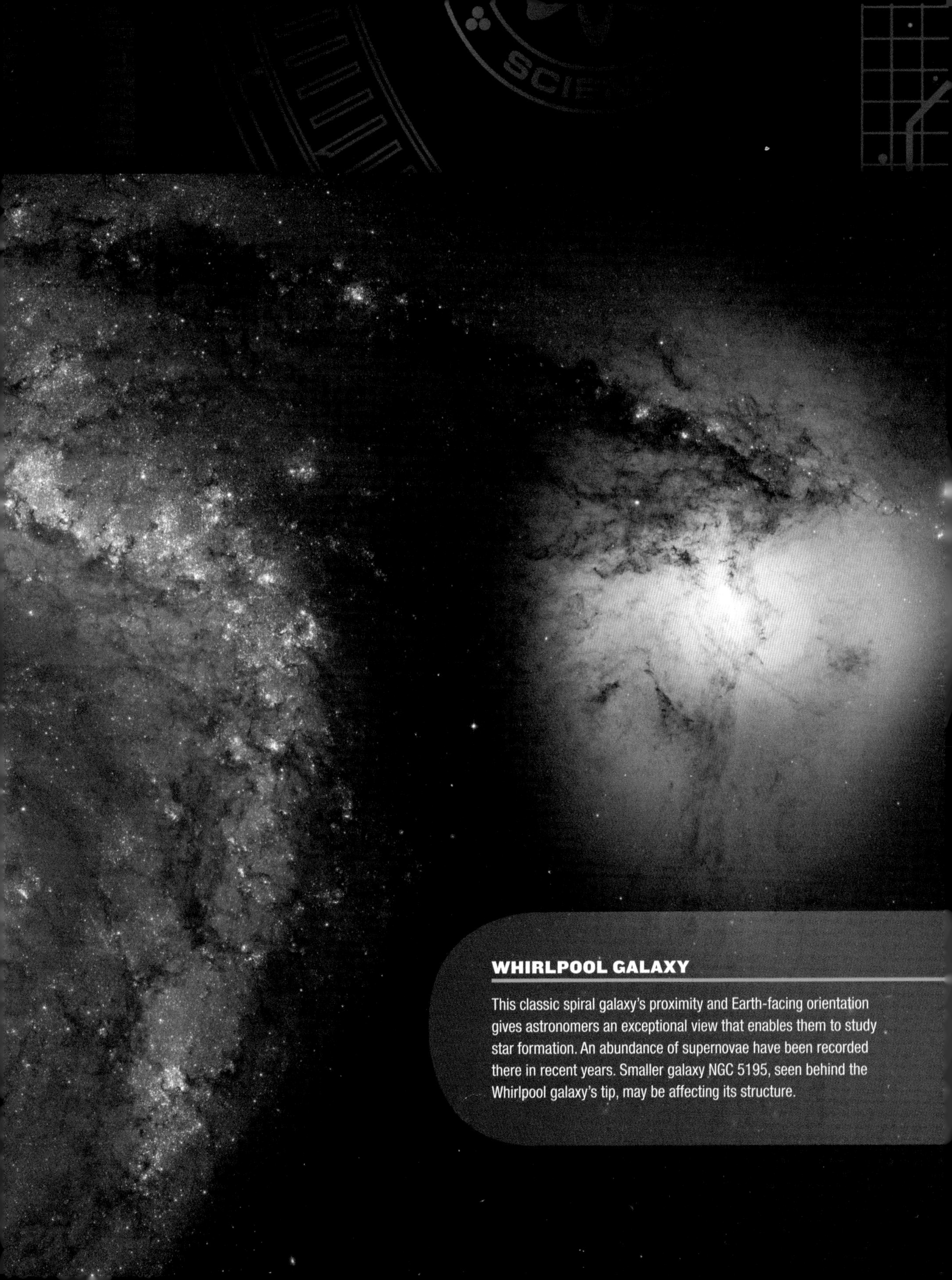

WHIRLPOOL GALAXY

This classic spiral galaxy's proximity and Earth-facing orientation gives astronomers an exceptional view that enables them to study star formation. An abundance of supernovae have been recorded there in recent years. Smaller galaxy NGC 5195, seen behind the Whirlpool galaxy's tip, may be affecting its structure.

SPIRAL GALAXY IN THE NIGHT SKY

Look for the Andromeda galaxy, within the constellation Andromeda.

STARGAZING TIPS

BEST VIEWING SPOTS: Under dark skies in the Northern Hemisphere and Southern Hemisphere down to the 35th parallel, including the southern part of South America, Australia, and Africa

BEST TIME TO SEE IT: Evening, September to March

BASIC TIPS: The Andromeda galaxy, or Messier 31, is a beautiful spiral galaxy that contains 300 billion stars and stretches 150,000 light-years across, making it larger than our own Milky Way. Sitting within its namesake constellation about 2.6 million light-years away, Messier 31 is one of the farthest objects visible to the unaided eye. It shines at magnitude +3.4 and can be spotted with the naked eye as a faint elliptical glow in the overhead skies of autumn during moonless nights.

HOW TO FIND IT

1. Look high in the south for the landmark asterism the Great Square, which marks the large constellation Pegasus, the Flying Horse. Two chains of stars appear to shoot out from the upper left corner of the square marked by 2nd magnitude star Alpheratz.

2. Draw a line from Alpheratz to faint naked-eye star Delta Andromedae and to brighter Mirach. Then hop to fainter +3.9 magnitude Mu Andromedae, and again hop the same distance—about the width of three middle fingers held at arm's length—to reach M31, a small elongated haze. Binoculars can aid the star hunt under light-polluted suburban skies.

3. A second route to find M31 starts with the bright W-shaped constellation Cassiopeia, the Queen. Use the three 2nd magnitude stars that make the right side of the "W" as an arrowhead that points directly to the Andromeda galaxy. The span between the arrowhead and M31 is equal to about three times the height of Cassiopeia's giant W formation. Binoculars will reveal Andromeda as an elongated cloud extending 3 degrees into the sky—equal to six moon disks. Telescopes will begin to reveal it as a bright oval patch of light that is the galaxy's star-packed core.

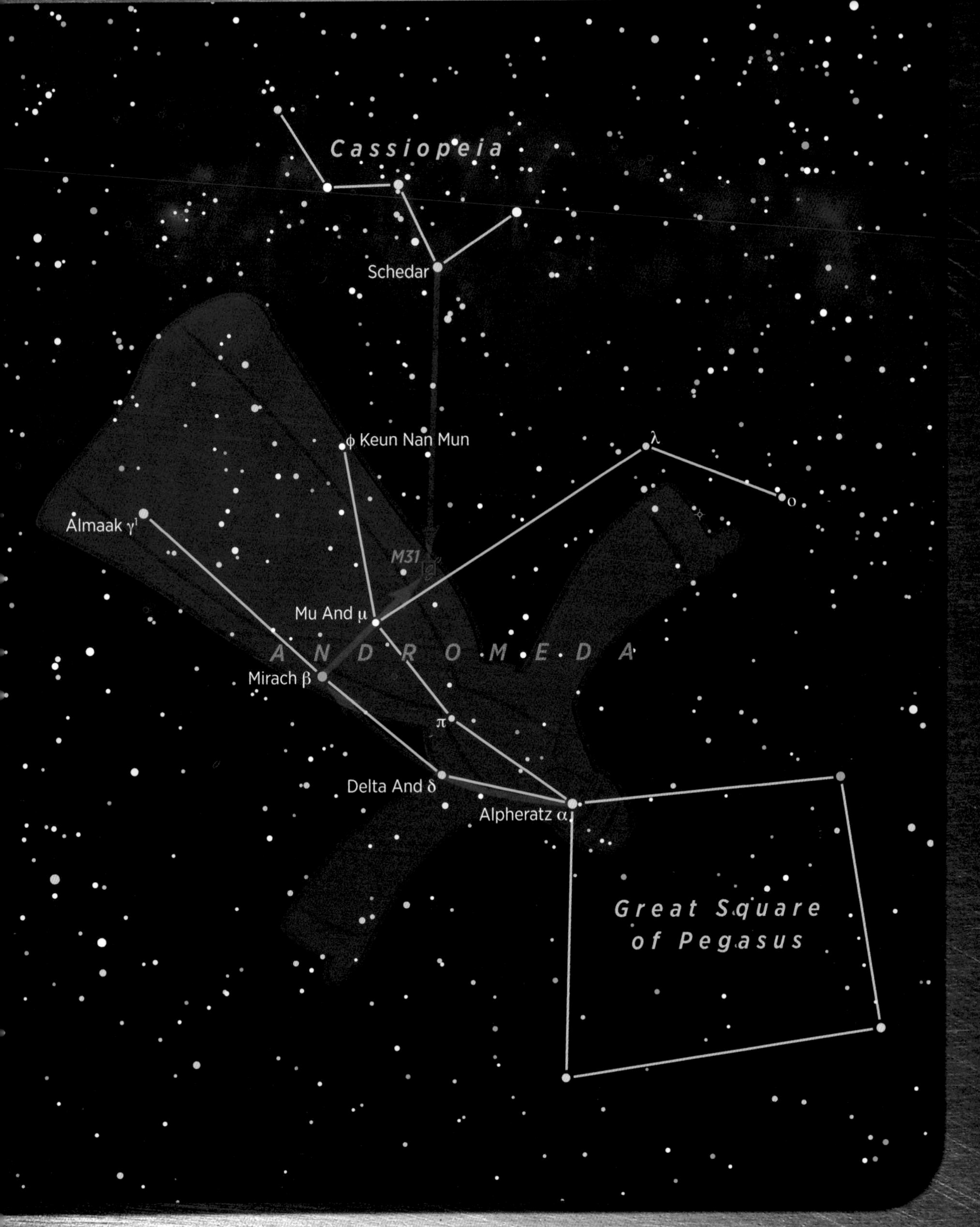
Cassiopeia
Schedar
φ Keun Nan Mun
λ
ο
Almaak γ¹
M31
Mu And μ
ANDROMEDA
Mirach β
π
Delta And δ
Alpheratz α
Great Square
of Pegasus

"Strange. Step-by-step, I've made the correct and logical decisions, and yet two men have died."
—SPOCK
NCC-1701/7

QUASARS IN *STAR TREK*

TOS / "THE GALILEO SEVEN"
SEASON 1 / 1967

IN THIS EPISODE: En route to deliver medical supplies to Makus III, the *U.S.S. Enterprise* veers off course to study the quasar Murasaki 312. Captain Kirk sends Spock and some of the crew out in the shuttlecraft *Galileo* to explore this "electromagnetic phenomenon," but they find it treacherous to navigate. They lose control of the ship and are forced to crash-land on lonely, inhospitable Taurus II. As Captain Kirk and the crew on board the *Enterprise* try to locate and save the *Galileo,* Spock grapples with command and the ways in which the situation challenges his devotion to logic.

THE QUASAR MURASAKI 312 STRETCHES OUT IN A tight green disk with a white-green rod of light appearing to spear its center. It moves outward in a blooming green cloud, or, as Kirk puts it, the quasar "whirls like some angry blight in space." The crew ventures toward it per Kirk's standing orders: Anytime they approach a quasar, they are duty bound to explore its phenomena. But it proves more dangerous than anticipated: As Spock moves the *Galileo* toward the quasar, they find their readings become erratic. Quasars in *Star Trek* are known to fluctuate and cause erratic shifts, although such volatility wouldn't necessarily be the case in our universe.

The crew is surprised when radiation levels spike and they cannot message back to the *Enterprise*. That's because Murasaki 312 ionized a whole sector of space, creating something called the "Murasaki Effect." This effect is what hides Taurus II from sensors, prompting Kirk to lament that compared to locating his lost crewmembers inside the quasar, "finding a needle in a haystack would be child's play."

Spock commands class F shuttlecraft Galileo *as it explores a quasar, Murasaki 312, that proves more treacherous than the crew imagined.*

A SHUTTLE REBORN

The Galileo *is destroyed in "The Galileo Seven" (TOS), but it still manages to pop up in four more episodes before it's painted to read* Galileo II.

Despite the crew's imperative to explore quasars when encountered, these cosmic powerhouses remain a mystery in the *Star Trek* universe. Tuvok tries to explain the Kelemane's planet by saying it rotates quickly "like a quasar" (VGR / "Blink of an Eye"), but explanations as to what drives quasars' behavior remain elusive. In whatever ways *Star Trek* quasars differ from those in the real world, they are equally as intriguing.

QUASARS IN OUR UNIVERSE

**DISCOVERED IN THE 1950S
CAN BE A TRILLION TIMES BRIGHTER THAN THE SUN / MORE THAN 2,000 KNOWN TO EXIST**

Nestled at the heart of the largest galaxies is a supermassive black hole with the mass of millions, or even billions, of sunlike stars. On the larger end of that spectrum, supermassive black holes containing a billion solar masses may have the power to produce the brightest objects in the universe: quasars.

QUASARS, OR OTHER FORMS OF "ACTIVE GALACTIC nuclei," were common in the early universe but are rare now. The ones we can see are very bright, energetic, and impressive in the extreme, but they are relics from a distant past. We don't have quasars nearby to study: The closest to us is 3C 273, about 2.5 billion light-years away—which is a thousand times farther away than the Andromeda galaxy. Thus the mandate for Kirk and the *U.S.S. Enterprise* to study any quasars they find is futile; the only way they would actually encounter one is to be transported to another galaxy. These phenomena could be studied remotely, of course, but that wouldn't be nearly as exciting as piloting a space shuttle right into the heart of one.

SEEING THE LIGHT

Quasars are the result of a galaxy's supermassive black hole throwing off energy while it absorbs matter. These monsters shine with the light of 10 to 10,000 times the brightness of the entire Milky Way—in fact, quasars outshine any galaxy that hosts them. Quasars are also highly variable, with fluctuations in brightness that occur over the course of a few hours. For changes to occur that swiftly, they must be small objects, smaller in diameter than our solar system.

In addition to their incredible brightness and small scale, another distinguishing feature of quasars is their jets. Particles spew outward from opposite poles at the center of the galaxy. Viewed from the side, quasars can be seen for billions of light-years.

COSMIC LASER BEAMS

When quasars aim their jets our way, they're known as blazars. The relativistic jets are composed of particles and radiation, generated when matter is pulled toward the supermassive black hole. A small fraction of that matter is accelerated away in a jet. Though the jet material may appear to be moving faster than the speed of light, it's an illusion; in reality, the jets move at about 99 percent of light speed. These jets remain well

ARE WE THERE YET?

HOW LONG CAN WE LAST?

In *Star Trek,* suspended animation means shutting down life processes and starting them up again when it's convenient to do so. Our real-life forays into suspended animation aren't about space travel: They're for treating trauma. In 2014, doctors in Pittsburgh started trials—cooling people, draining their blood, and replacing it with a solution of saline—in hopes of giving gunshot wound patients time.

Getting There

TOS / "Space Seed"

Quasars, like the one in this artist's depiction, are at the center of some galaxies. These monsters are powered by a supermassive black hole, and their relativistic jets are capable of accelerating matter to near light speed.

collimated—very focused, in other words, like a laser beam—for millions of light-years.

Quasars and blazars are so bright that it is easy to see them at great distances, so a starship doesn't need to wait until it approaches one to study it. Real 21st-century technology could support quasar observation from anywhere with an unobstructed view of the target, posing little to no risk to a starship, its equipment, or its crew.

COSMIC FLASHLIGHT

Quasars emit so much energy in the form of electron volts that they shine 10 to 100,000 times brighter than the Milky Way.

BEHIND THE SCENES

In building the world of the future, *Star Trek*'s creators had the difficult task of dreaming up unique alien species and bringing them to life. Through extensive makeup, face-changing prosthetics, and careful costuming, they gave us the distinctive forehead ridges of the Klingons, the prominent ears of the Ferengi, and a rainbow of out-of-this-world skin colors. These sketches and behind-the-scenes photos show how *Star Trek*'s costume designers and makeup artists made these aliens part of our world.

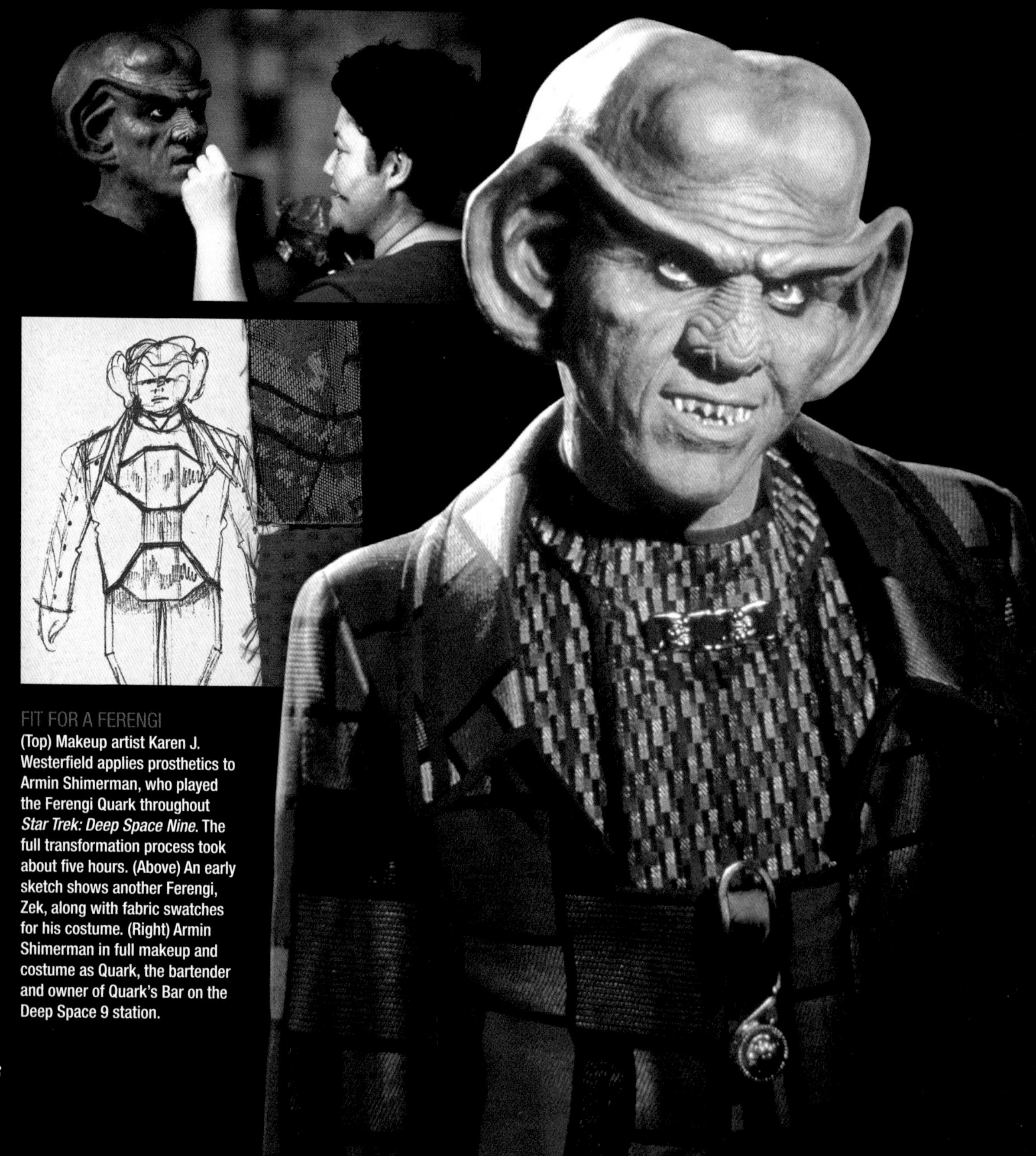

FIT FOR A FERENGI

(Top) Makeup artist Karen J. Westerfield applies prosthetics to Armin Shimerman, who played the Ferengi Quark throughout *Star Trek: Deep Space Nine.* The full transformation process took about five hours. (Above) An early sketch shows another Ferengi, Zek, along with fabric swatches for his costume. (Right) Armin Shimerman in full makeup and costume as Quark, the bartender and owner of Quark's Bar on the Deep Space 9 station.

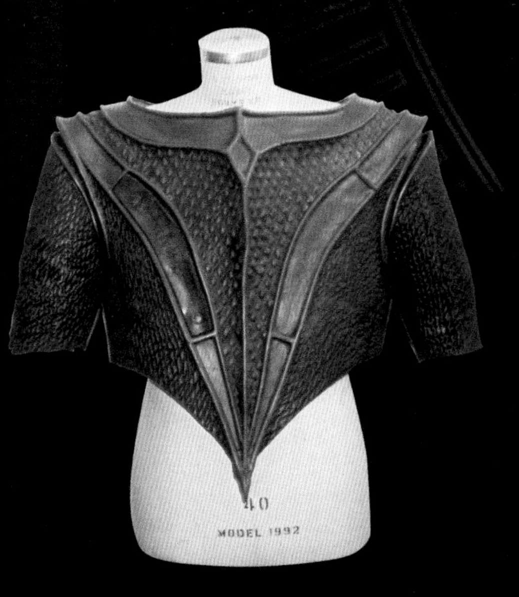

KEEPING UP WITH THE CARDASSIANS (Top and left) Costume designer Robert Blackman created this heavy, armorlike uniform for the Cardassians. Initially created for *Deep Space Nine*, the same costumes were used for Cardassian characters on *The Next Generation.* (Above) Casey Biggs as Damar, leader of the Cardassian Rebellion, is seen in uniform.

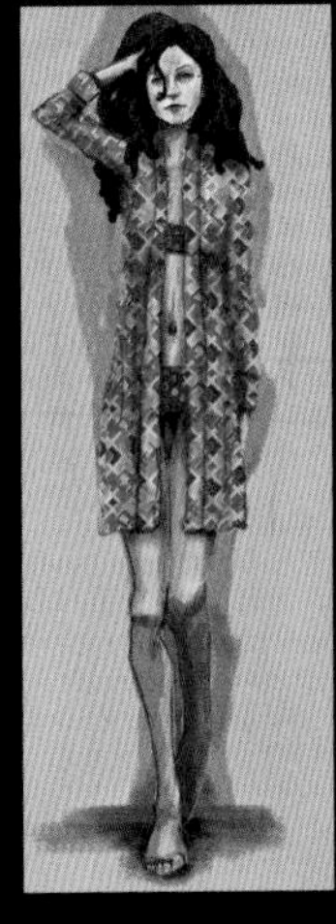

AN UPDATED ORION Character art from *Star Trek* (2009) shows one of the film's costumes for Gaila, an Orion and a cadet at Starfleet Academy. Traditionally Orions have black hair, but makeup artists for the film decided to give Gaila, played by Rachel Nichols, red hair instead.

THE BORG IDENTITY Artwork shows the early concept for the Borg costumes (left) and the finished product (below). Each member of the Borg Collective had a unique costume, with mechanical pieces replacing or enhancing worn-out human parts.

QUASAR IN THE NIGHT SKY

Look for 3C 273, a quasi-stellar object within the constellation Virgo.

STARGAZING TIPS

BEST VIEWING SPOTS: Across the Northern Hemisphere during spring and summer, and in the Southern Hemisphere autumn and winter

BEST TIME TO SEE IT: Late evenings, April to July

BASIC TIPS: Located 2.4 billion light-years from Earth, 3C 273 is tucked away in the springtime constellation Virgo, the Maiden, and looks like an unassuming faint blue-tinted starlike object. It's not easy to find and observe quasars in the night sky. Even with 3C 273 being the brightest of its kind, visible at magnitude +12.9, it still takes a suburban telescope with at least an 8-inch mirror to glimpse. The joy of observing 3C 273 comes not as much from seeing it, but from understanding that what we are seeing is probably the farthest object in the universe that most backyard astronomers will ever observe.

HOW TO FIND IT

1. Look high in the northeast sky for an upside down Big Dipper. Draw an imaginary line from the dipper's handle stars to the bright orange star Arcturus. Extend the line to the next superbright star Spica, lead member in the constellation Virgo, the Maiden.

2. From Spica, scan over to much fainter +2.7 magnitude star Porrima, or Gamma Virginis. The two stars are separated by about 15 degrees, which is equal to the tip-to-tip wide span of the index finger and little finger held at arm's length.

3. Sweep 5 degrees from Porrima to Zaniah, equal to about the width of three middle fingers held at arm's length. The faint starlike quasar shines with a distinct blue color, sitting above this star pair to form an equilateral triangle pattern.

Big Dipper
Alkaid
Arcturus
Vindemiatrix ε
VIRGO
ο
ν
δ Auva
3C 273
β Zavijava
τ
Porrima γ
η Zaniah
ζ Heze
μ Rijl al Awwa
θ
ι Syrma
Spica α

NAVIGATING THE NIGHT SKY

LONG BEFORE THE FIRST FEDERATION SHIPS BROKE the warp barrier and began sailing to distant stars across the galaxy, our view of the universe was the same as that seen by the first Terran astronomer looking up at the night sky—a perspective seen only from planet Earth. And even after centuries of leaving the bounds of our homeworld and its solar system, many Starfleet cadets surely trace their fascination with the cosmos to earlier times spent gazing upward at the star-filled heavens—just like us, their forebear astronauts from the 21st century.

THE CELESTIAL SPHERE

All those pinpoints of light above are part of a giant celestial sphere, half of which is hidden from our view by the sun's glare. Earth itself also blocks the view so that people living in the Northern Hemisphere see mostly the northern half of the celestial sphere, and those in the Southern Hemisphere are best positioned to see the southern counterpart. Earth spins around on its axis, which runs through an imaginary line from the North Pole to the South Pole, once every 24 hours. This spin creates the effect of having all celestial objects appear to rise in the east, glide across the southern sky, and set in the west.

At a glance, it may seem an unchanging backdrop, but what we can see of the heavens changes with the viewer's location on Earth, the time of day (or night), and with the season, in a yearly cycle. Observers at any location on Earth will always see stars rising in the east and setting in the west. However, the visibility of a constellation's stars on any one night and its path across the sky depend on the sky watcher's latitude as our spherical planet spins on its axis. For example, an observer located at the North Pole will have the north celestial pole located directly overhead—pinned on Polaris, or the North Star—while all other stars appear to glide counterclockwise around it throughout the night. These stars, visible in the northern sky and appearing never to set, are called circumpolar stars. In the Southern Hemisphere, the different stars close to the southern celestial pole are also circumpolar, moving east to west, but appear to circle clockwise. (Try it for yourself with a spinning ball.)

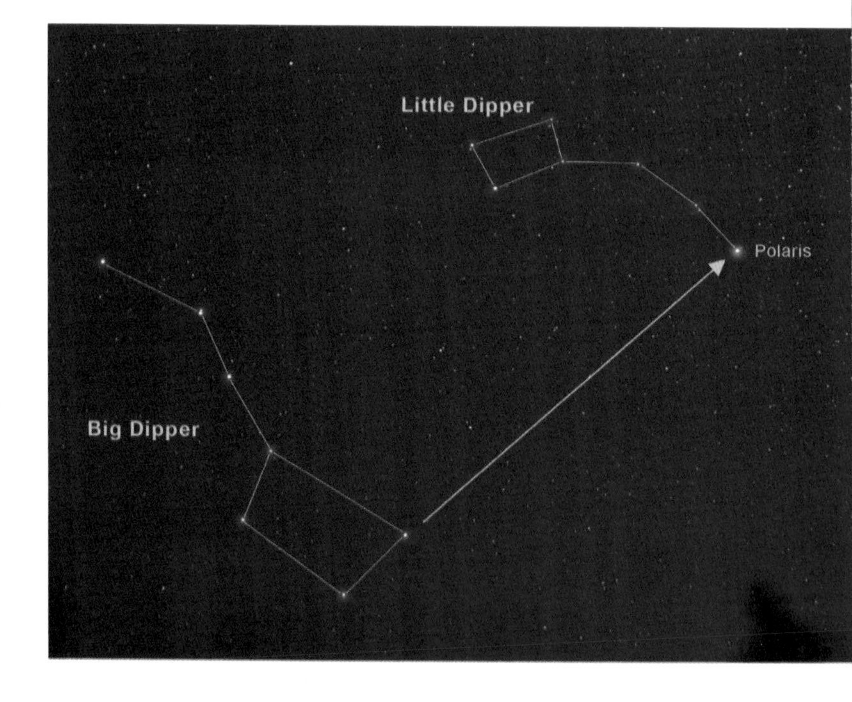

To locate the north celestial pole, start with a recognizable constellation: the Big Dipper. Draw an imaginary line straight out of the dipper's bowl until you reach a bright star—Polaris, also known as the North Star.

For an observer on the Equator, the stars appear to rise from the east, move overhead, and set to the west.

As Earth orbits around the sun, the night side of our planet faces different constellations during the course of a year in a progression that means stars appear to rise four minutes earlier each night. Then, about one year later, the cycle starts again. This means that the appearance over the southern horizon of a certain star, say Sirius, the Dog Star, in the mid-northern latitudes, means that the "dog days" of summer have arrived.

Since ancient times, people have imagined stars projected on this celestial sphere and grouped them into constellations. This is an illusion, because the stars forming any given constellation lie at varying distances from us and each other. So the patterns we see them form in the night sky happen purely due to our viewpoint from Earth.

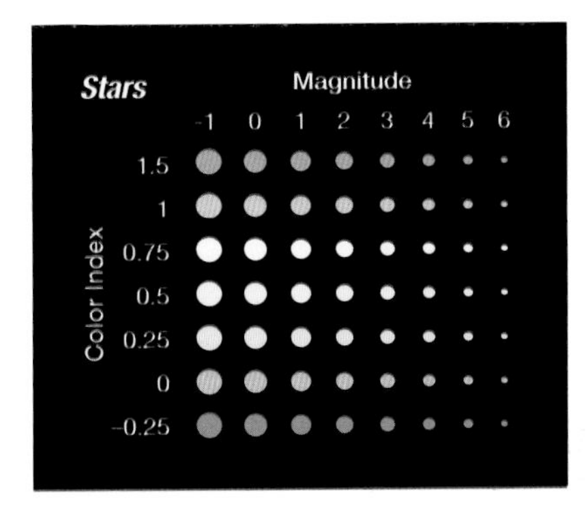

Stars come in all different colors and levels of brightness. Blue stars burn at the highest temperatures, and red the lowest. The bigger and brighter a star is, the lower its apparent magnitude.

STAR BRIGHTNESS

Even with your naked eye you can see that not all stars are the same brightness. The ancient Greek astronomer Hipparchus probably created the first stellar catalog, dividing stars into different classes of brightness or magnitude and establishing a system that is still in use today. This scale refers to a star's brightness as it appears in the night sky or what is called "apparent magnitude." The brighter a star appears in the sky, the lower its magnitude number, with the brightest objects even having negative numbers. Fainter stars have larger magnitude numbers. The scale has been extended to include even fainter objects visible only through telescopes. Star charts indicate the magnitude of stars with a graded series of dot sizes. The fainter a star in the sky, the tinier its counterpart dot on a star chart, with each size representing an order of magnitude in brightness.

VISION LIMITS

Sky watchers stuck within city suburbs can see stars down to around magnitude 4 on a clear moonless night. From a pristine dark location, far from any city lights, the unaided human eye can see stars down to magnitude 6. However, binoculars will reveal stars as faint as magnitude 8 or 9 while a small backyard telescope can reach down to magnitude 12. Meanwhile, the Hubble Space Telescope in orbit above the Earth can record objects as faint as magnitude 31. This is about two billion times fainter than the faintest object visible to the naked eye.

A star's brightness in the sky depends on its size, temperature, and distance from us. More massive stars tend to burn brighter than smaller ones. But also the closer a star is to Earth, the brighter it tends to appear. So although brightness or magnitude is an important characteristic, it is relative, and based on an earthbound perspective.

COSMIC HEADINGS

To find their way around the heavens, novice stargazers first need to know how to find north and south. Luckily for those in the Northern Hemisphere, a moderately bright star, Polaris, lies close to the north celestial pole. To track it down, start with the Big Dipper in the constellation Ursa Major, the Great Bear, that lies near the horizon in autumn and rides high in the sky in spring for mid-latitude locales. Draw an imaginary line between the two stars at the end of the dipper's bowl and extend it straight out another 25 degrees—about the length of the entire Big Dipper—until you reach a bright star: Polaris, also called the North Star.

Observers in the Southern Hemisphere need to use a bit more indirect method for identifying the location of the south celestial pole. Because there is no bright star marking it, start by finding the star pattern known as the Crux, the Southern Cross. Draw a line along the cross's long axis and a gentle arc out beyond it to the bright star Achernar. The pole lies nearly halfway between—about 26 degrees from Crux, equal to the span from the tip of your little finger and your thumb when stretched as far apart as possible, held at arm's length.

The apparent sizes of objects and the distances that separate them in the sky are measured in angles, expressed in degrees, minutes, and seconds. For example, the distance (or angle) from the horizon to a point directly overhead is 90 degrees. But it can be tricky to translate these measurements from handheld star charts directly to the real sky above. An easy stargazing trick is to use your own hands and fingers held at arm's length and the famous Big Dipper star pattern in the northern sky as a convenient, portable angle measurer.

HANDS UP

Your outstretched hand is about 25 degrees wide from the tip of the thumb to the tip of the little finger. This is roughly equal to the length of the Big Dipper: the distance between the last star in the handle and the end stars in the bowl. Smaller distances can be measured with your fist, which is about 10 degrees across and is about equal to the width of the dipper's bowl itself. Three middle fingers measure about 5 degrees across, or about the same as the depth of the dipper's bowl. For smaller sky angles use your thumb, which is equal to 2 degrees, and finally your index finger, about half a degree, which can easily cover up either the sun or the moon.

STAR HOPPING

The technique of star hopping is a popular way for both novice and seasoned backyard sky watchers to navigate the night sky. It involves using easily visible bright stars as jumping-off points to locate dimmer celestial objects, and it's especially important if you are viewing—as many of our Stargazing pages suggest you do—a region of the sky where a very distant object is located.

Start by using the star chart or stargazing apps like Star Walk and SkySafari on your mobile device to identify a pattern of stars that you can easily recognize in the sky, like the Big Dipper, Orion's belt, or the Teapot. Instructions can guide you from the pattern toward another nearby bright star and, using the geometry among them, lead you in hopping to other stars—some perhaps fainter—until you arrive at your target's location. Binoculars will help you locate many faint deep-sky objects. As you become more comfortable cruising among the constellations and locating ever fainter deep-sky challenges, star maps either printed or digital will remain your key to charting the course among the starry skies.

UNDERSTANDING SKY MAPS

The following pages showcase a series of four maps of the entire night sky for the Northern Hemisphere, one for each season. Each map depicts what the evening starry sky looks like for an observer located at 40° north latitude, so it is centered for those in United States, Canada, Europe, and Japan, but is generally

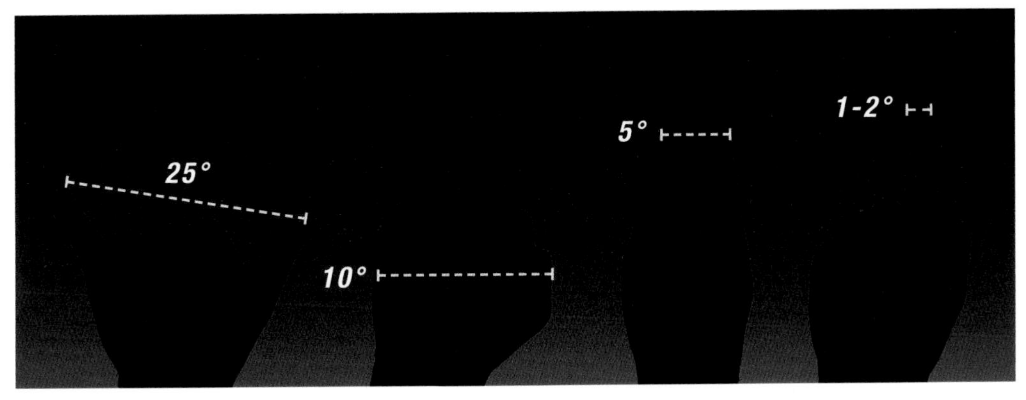

In a pinch, your hand can serve as a tool to estimate different angles in the sky. Pinky to thumb is about 25 degrees, a fist is about 10 degrees, the three middle fingers together are about 5 degrees, and a single finger is about 1 degree.

usable 20° north and south of this latitude. The North Star (Polaris) and northern constellations like Ursa Major and its Big Dipper will appear higher in the sky for locations north of 40° north latitude and lower in the sky for stargazers farther south, closer to the Equator. Each map represents the sky on specific dates and times, as indicated, but can serve as a general guide to the skies of that season.

MAP FEATURES

Constellations are marked with stars down to magnitude 5, and stars that are magnitude 3.5 or higher are named and tinted with their color. A selection of the most famous deep-sky objects, those beyond the solar system, like star clusters, nebulae, and galaxies, are marked with a unique set of symbols. Each map includes a dotted line representing the ecliptic—the pathway of the sun—near which the moon and planets can also be found. The faint white band in the background marks the disk of the Milky Way.

BASIC STARGAZING EQUIPMENT

Binoculars should be your first equipment choice. They give you a wide field of view, so you can easily sweep the sky and hunt down cool objects like star clusters. They are also fairly inexpensive, widely available, and a cinch to carry around.

If you find yourself seriously committed to studying the night sky, it's time to graduate to a telescope. When it comes to telescopes, size really does matter. The wider the main mirror or lenses inside the telescope tube, the more light you can capture, which means your views will be sharper and brighter. The best all-around starter scopes that are easy to use for the entire family should have mirrors four to six inches wide and a rock-steady mount.

There's one other piece of equipment that every beginning stargazer will want. Think about it: How do you read star charts in the dark? You devise a red-colored light. That's because red light has the least effect on night vision. You can purchase red-light flashlights at an outdoor store, or you can make one by securing red cellophane with a rubber band over the front of a regular flashlight. Remember that our eyes need at least 15 minutes to adapt to seeing faint things in the darkness, so give yourself plenty of time when you go out to look at the stars. Don't forget to turn on the red "night sky" modes in stargazing apps and turn off your device's blinding flashlight modes so you can retain your night vision.

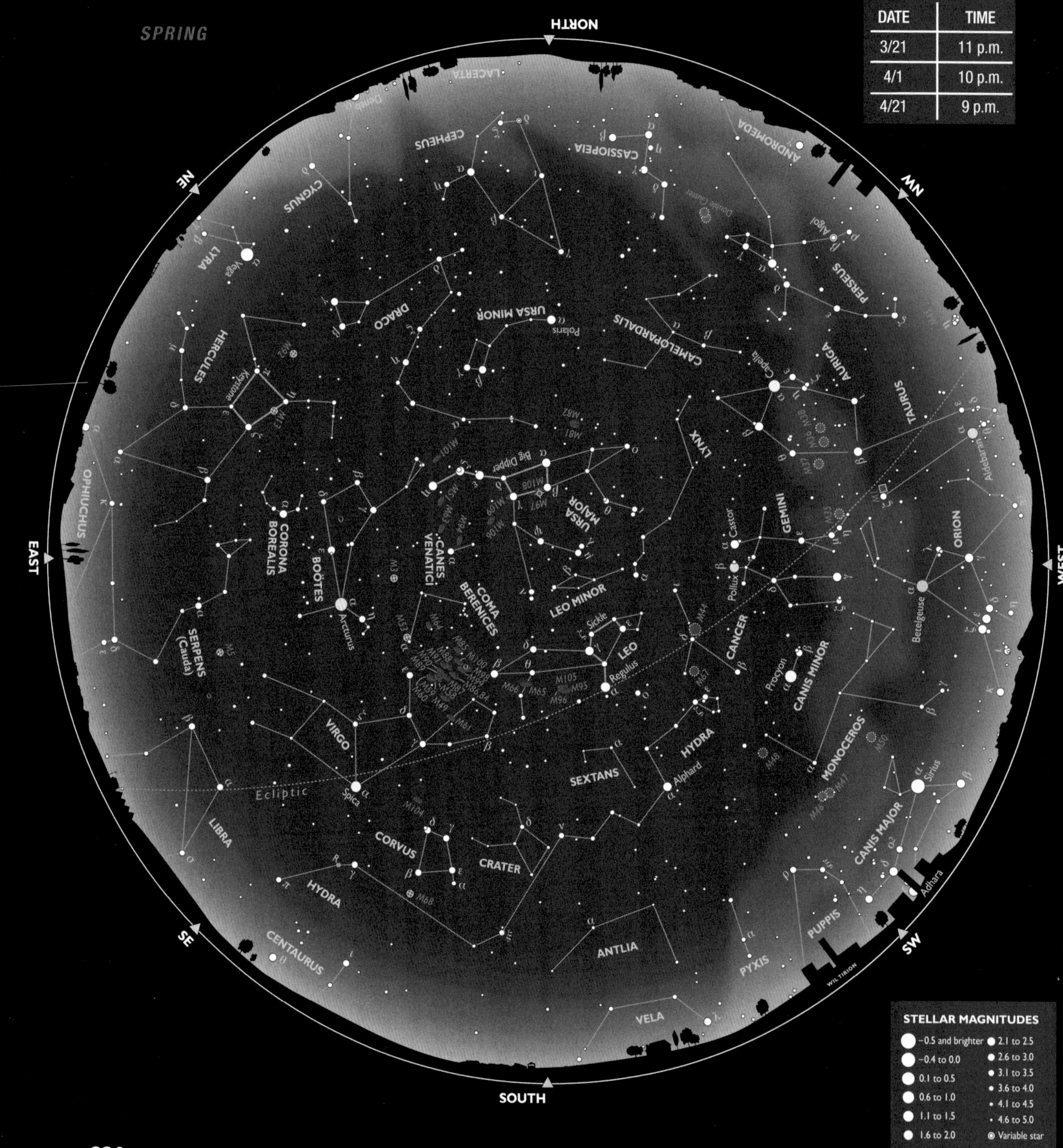

DATE	TIME
3/21	11 p.m.
4/1	10 p.m.
4/21	9 p.m.

STELLAR MAGNITUDES

●	–0.5 and brighter	●	2.1 to 2.5
●	–0.4 to 0.0	●	2.6 to 3.0
●	0.1 to 0.5	●	3.1 to 3.5
●	0.6 to 1.0	●	3.6 to 4.0
●	1.1 to 1.5	●	4.1 to 4.5
●	1.6 to 2.0	●	4.6 to 5.0
		⊙	Variable star

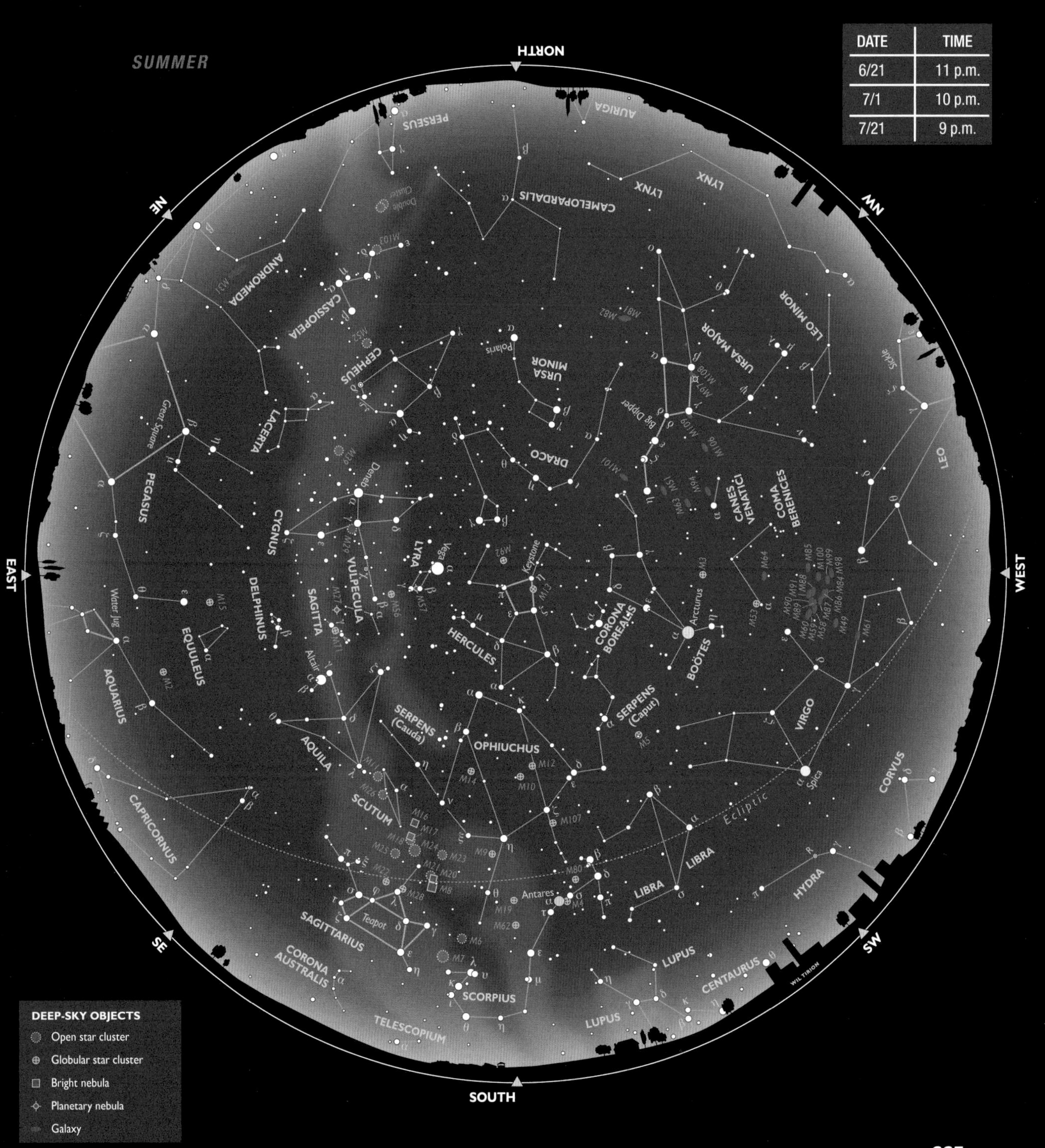

DATE	TIME
6/21	11 p.m.
7/1	10 p.m.
7/21	9 p.m.

DEEP-SKY OBJECTS
- Open star cluster
- Globular star cluster
- Bright nebula
- Planetary nebula
- Galaxy

NIGHT SKY CHARTS

AUTUMN

DATE	TIME
9/21	11 p.m.
10/21	10 p.m.
11/1	8 p.m.

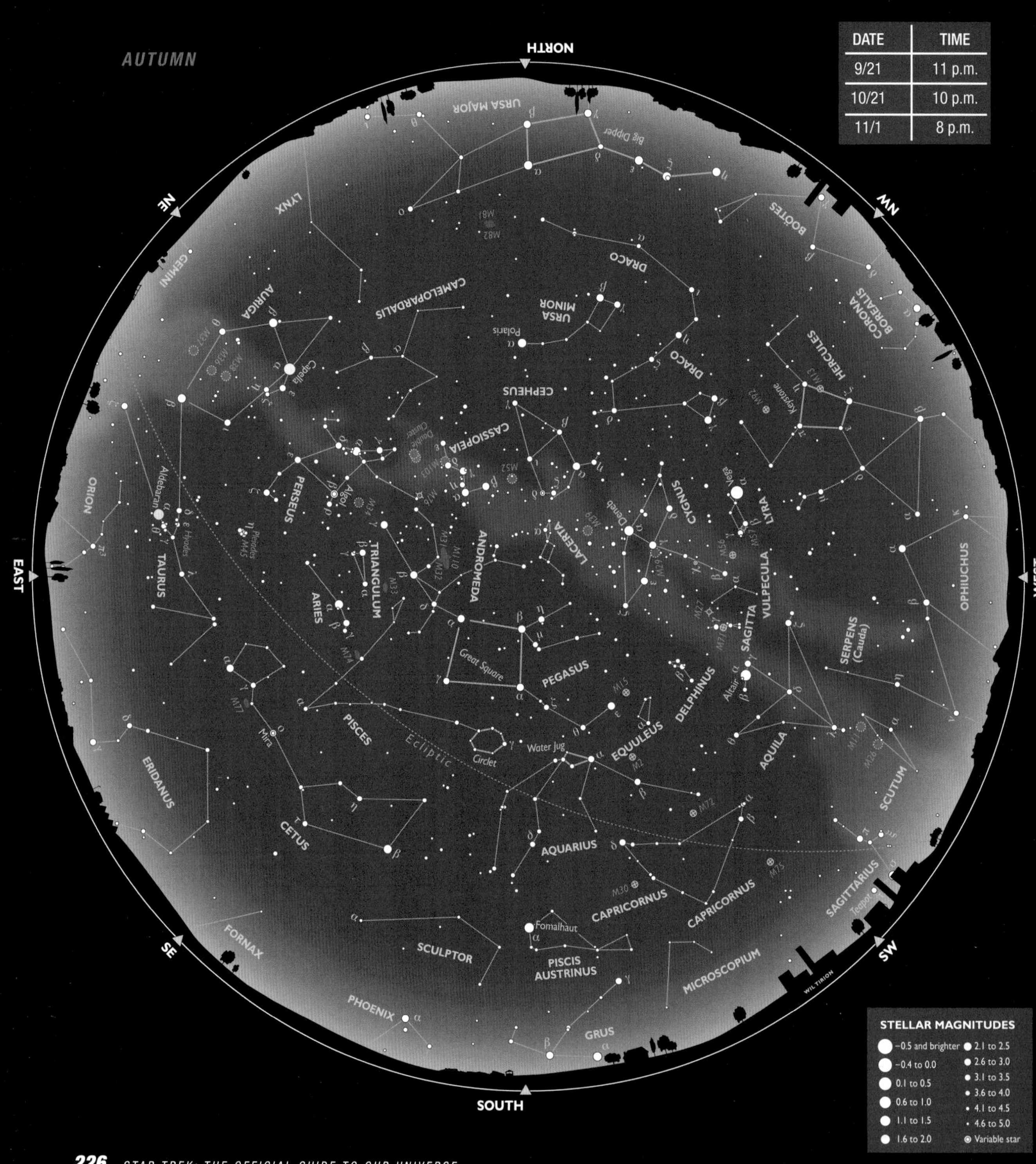

WINTER

DATE	TIME
12/21	11 p.m.
1/21	9 p.m.
2/1	8 p.m.

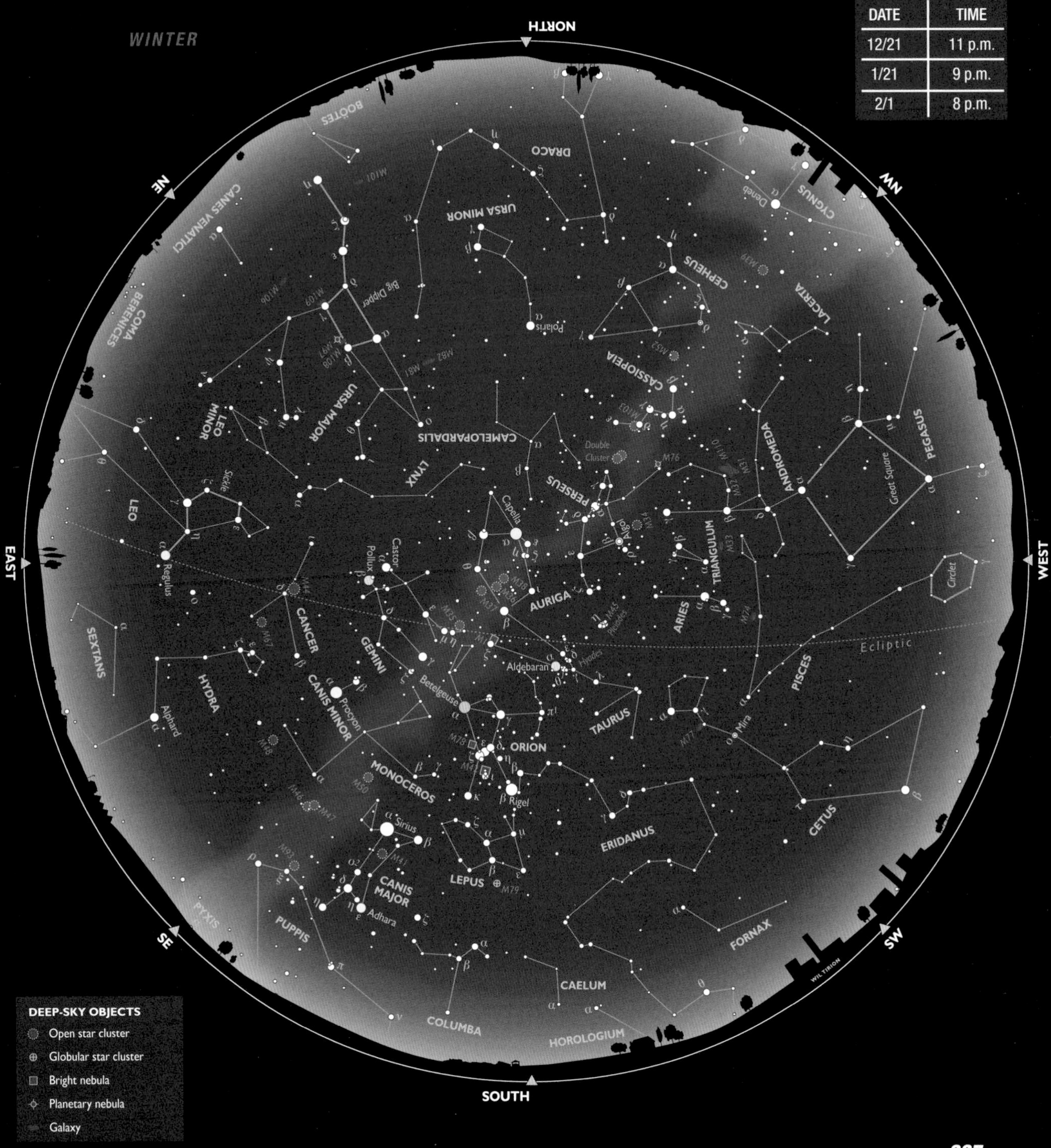

DEEP-SKY OBJECTS

- Open star cluster
- Globular star cluster
- Bright nebula
- Planetary nebula
- Galaxy

ACKNOWLEDGMENTS

LIKE THE MANY STARSHIP VOYAGES WITHIN THE vast *Star Trek* universe, the realization of this beautiful book has been a cosmic adventure that could only have been accomplished by an intrepid team, which I am humbled and honored to have had at National Geographic Books.

I would like to give an everlasting Vulcan salute to senior editor Susan Hitchcock, whose expert guidance and eternal enthusiasm from the beginning made this monumental and complex project possible. My heartfelt thanks also goes to project editor Anne Smyth, senior editor Robin Terry-Brown, and assistant editor Michelle Cassidy for patiently navigating and juggling the milestones involved in weaving together the myriad *Star Trek* and real science components.

I also want to thank director of photography Susan Blair and photo editor Patrick Bagley for skillfully pulling together a stellar collection of astronomical and *Star Trek* photos, and Debbie Gibbons, director of intracompany and custom cartography, and senior graphics editor Matt Chwastyk for their stunning star maps. Special appreciation goes to Marty Ittner, our art director and designer for the jaw-dropping layouts and designs that give the book its Starfleet appeal.

To science researcher Kate Carroll, *Star Trek* text editor Kate Armstrong, and astronomy consultant Beth Hufnagel, your attention to details, creativity, and investigating skills know no bounds; I am eternally indebted.

Gratitude goes to the staff at CBS Consumer Products for sharing their *Star Trek* wisdom. We are particularly appreciative of John Van Citters, vice president of product development, for assisting in vetting and reviewing all aspects of *Star Trek* content, as well as Marian Cordry and Risa Kessler for opening the vast vault of amazing *Star Trek* imagery and helping us choose rarely seen photos from the many iconic series and movies.

The entire team also sends their thanks to the legions of dedicated *Star Trek* fans who were integral in the creation of the phenomenal encyclopedic website Memory Alpha, which served as a wonderful resource for background information to the franchise.

And, of course, our thanks goes to Gene Roddenberry for creating this fantastical universe we have all gotten to share in this project.

Finally, this book was such an epic undertaking that I could not have completed it without the moral support and eternal patience of my wife Zoe, two daughters Sophie and Kathleen, and the continuous encouragement of my parents Arpad and Barbara.

EPISODE INDEX

Throughout this book, you will find references to specific moments in the *Star Trek* television shows and movies. These references have been gathered here in a handy index, which is organized chronologically by series.

Cast of Star Trek, *season 3 (1968–69)*

THE ORIGINAL SERIES (TOS)

1966–1969

THE ANIMATED SERIES (TAS)

1973–1974

Cast of Star Trek: The Next Generation, *season 6 (1992–93)*

THE NEXT GENERATION (TNG)

1987–1994

DEEP SPACE NINE (DS9)

1993–1999

Cast of Star Trek: Deep Space Nine, *season 7 (1998–99)*

Cast of Star Trek: Voyager, *season 4 (1997–98)*

VOYAGER (VGR)

1995–2001

Cast of Star Trek: Enterprise, *season 3 (2003–04)*

ENTERPRISE (ENT)

2001–2005

ALTERNATE REALITY MOVIES

2009–

Cast of Star Trek *(2009)*

INDEX

Boldface indicates illustrations.

ILLUSTRATIONS CREDITS

All images provided by CBS Studios, Inc., with the exception of the following:

Cover, composite image by Jonathan Halling; cover (Orion Nebula), NASA, ESA, M. Robberto (Space Telescope Science Institute/ESA) and the Hubble Space Telescope Orion Treasury Project Team; back cover (Andrew Fazekas), Tristan Brand; back cover (William Shatner), Rory Lewis; 11, J. S. Urquhart modified from the original presented in Urquhart et al., 2014, MNRAS, 437, 1791, background artwork by Robert Hurt, the Spitzer Science Center; 16-7, Sophie Jacopin/Science Source; 18-9, Moonrunner Design/NG Maps; 23, Dana Berry/National Geographic Creative; 31, Art by Stefan Morrell. Sources: Christopher McKay. NASA Ames Research Center; James Graham, University of Wisconsin-Madison; Robert Zubrin, Mars Society; Margarita Marinova, California Institute of Technology. Originally published in *National Geographic* magazine, February 2010; 32-3, NASA/JPL-Caltech/MSSS; 39, NASA/JPL/DLR; 40-1, NASA/JPL/University of Arizona; 47, NASA/JPL; 55, Dan Schecter/Getty Images; 56, ESA/Rosetta/NAVCAM; 57 (UP LE), ESA/Rosetta/NAVCAM; 57 (LO LE), ESA/Rosetta/MPS for OSIRIS Team MPS/UPD/LAM/IAA/SSO/INTA/UPM/DASP/IDA; 57 (UP RT), ESA/Rosetta/NAVCAM; 62-3, ESO/L. Calçada; 64, Lynette Cook/Science Source; 65, Ethan Tweedie Photography; 68, NASA/Ames/JPL-Caltech; 69, art © Jon Lomberg, www.jonlomberg.com; 77, ESA, NASA, G. Tinetti (University College London, UK & ESA) and M. Kornmesser (ESA/Hubble); 78-9, ESO/L. Calçada; 84, Ken Murray/ZUMA Press/Corbis; 85, Graphic of the ten largest known trans-Neptunian Objects (TNOs) by Larry McNish, Calgary Centre of the RASC, updated 2016; 86-7, NASA/JPL-Caltech; 93, Eye of Science/Science Source; 100-1, ESO/M.-R. Cioni/VISTA Magellanic Cloud survey. Acknowledgment: Cambridge Astronomical Survey Unit; 102-3, www.bighistoryproject.com; 107, ESO; 108-9 (background), Irene Buil Mijaloff/Getty Images; 115, Royal Observatory, Edinburgh/Science Source; 116-7, Mark Garlick/Science Source; 122, Blue Origin; 123, John Sanford/Getty Images; 124-5, Alan Dyer/Stocktrek Images/Corbis; 131, NASA, ESA, CXC, SAO, the Hubble Heritage Team (STScI/AURA), and J. Hughes (Rutgers University); 132-3, ESO/IDA/Danish 1.5 m/R.Gendler, J-E. Ovaldsen, C. Thöne, and C. Feron; 133, ESA/Hubble & NASA; 138, Richard Kail/Science Source; 139, X-ray: NASA/CXC/Univ. of Toulouse/M.Bachetti et al, Optical: NOAO/AURA/NSF; 140 (LO), Photo of Bluetooth Communicator from the Wand Company; 146-7, NASA, ESA, and the Hubble Heritage Team (STScI/AURA)—ESA/Hubble Collaboration; 148, NASA, ESA, and the Hubble SM4 ERO Team; 149, NASA, H. Ford (JHU), G. Illingworth (UCSC/LO), M.Clampin (STScI), G. Hartig (STScI), the ACS Science Team, and ESA; 153, NASA, ESA, M. Robberto (Space Telescope Science Institute/ESA), the Hubble Space Telescope Orion Treasury Project Team and L. Ricci (ESO); 161, NASA, ESA/Hubble and the Hubble Heritage Team; 162-3, Author: Rafael Rodríguez Morales, amateur astrophotographer, www.skyandphoto.com. Captured on September 18, 2015—Seville, Spain; 168, Andrzej Wojcicki/Science Photo Library/Corbis; 169, Bill Snyder Astrophotography; 176, Jeremy Manning/Boeing; 177, NASA, ESA, and the Hubble Heritage (STScI/AURA)-ESA/Hubble Collaboration; 178-9, NASA, ESA, J. Hester, and A. Loll (Arizona State University); 184-5, NASA, ESA, and E. Sabbi (ESA/STScI); 186 (UP LE), NASA, ESA, and the Hubble Heritage Team (STScI/AURA); 186 (UP RT), NASA, ESA, and the Hubble Heritage (STScI/AURA)-ESA/Hubble Collaboration; 186 (CTR), NASA, ESA, and the Hubble Heritage Team (STScI/AURA); 186 (LO LE), NASA, ESA, K. Kuntz (JHU), F. Bresolin (University of Hawaii), J. Trauger (Jet Propulsion Lab), J. Mould (NOAO), Y.-H. Chu (University of Illinois, Urbana), and STScI; 186 (LO RT), NASA, ESA, and the Hubble Heritage Team (STScI/AURA); 187, NASA, ESA, and the Hubble Heritage Team (STScI/AURA)-ESA/Hubble Collaboration; 190, Paul Buck/epa/Corbis; 191, NASA, ESA, and AURA/Caltech; 198, TSGT. Douglas K. Lingefelt, U.S. Air Force/Science Source; 199, ESA/Hubble & NASA (acknowledgement: Gilles Chapdelaine); 200-1, NASA, ESA, and the Hubble SM4 ERO Team; 207, 2002 R. Gendler, photo by R. Gendler/www.spacetelescope.org; 208-9, NASA, ESA, S. Beckwith (STScI), and the Hubble Heritage Team (STScI/AURA); 215, ESO/M. Kornmesser; 223, Patrick J. Bagley/NG Staff; 239 (UP), Tristan Brand; 239 (LO), Rory Lewis.

ANDREW FAZEKAS, also known as the Night Sky Guy, is a science writer, speaker, and broadcaster who shares his passion for the wonders of the universe through all media. He is a columnist for National Geographic News, where he authors the popular online weekly StarStruck column. Fazekas is also a syndicated contributor for the Canadian Broadcasting Corporation radio network, the national science columnist for Yahoo News, and the communications manager for Astronomers Without Borders. As an active member of the Royal Astronomical Society of Canada he has given hundreds of public talks and educational workshops. Observing the heavens for more than three decades, he has never met a clear night sky he didn't like. He lives in Montreal, Canada. www.thenightskyguy.com

WILLIAM SHATNER's career spans more than 50 years as an award-winning actor, director, producer, writer, recording artist, and horseman. In 1966, he originated the role of *Star Trek's* Captain James T. Kirk, a role he played through three television seasons and in seven movies. Beyond those, he has acted in and directed hundreds of television shows, films, and documentaries. Shatner has also authored nearly 30 books, fiction and nonfiction, including an autobiography, *Up Till Now*—a *New York Times* bestseller—and, most recently, *Leonard: My Fifty-Year Friendship With a Remarkable Man.*

Since 1888, the National Geographic Society has funded more than 12,000 research, exploration, and preservation projects around the world. National Geographic Partners distributes a portion of the funds it receives from your purchase to National Geographic Society to support programs including the conservation of animals and their habitats.

National Geographic Partners, LLC
1145 17th Street NW
Washington, DC 20036

Become a member of National Geographic and activate your benefits today at natgeo.com/jointoday.

For information about special discounts for bulk purchases, please contact National Geographic Books Special Sales: ngspecsales@ngs.org

For rights or permissions inquiries, please contact National Geographic Books Subsidiary Rights: ngbookrights@ngs.org

LIBRARY OF CONGRESS CATALOGING-IN-PUBLICATION DATA

Fazekas, Andrew, author. | Shatner, William, author of foreword.
Star Trek, the official guide to our universe : the true science behind the starship voyages / Andrew Fazekas ; foreword by William Shatner.
Washington, D.C. : National Geographic Partners, LLC, [2016]
LCCN 2016007552 | ISBN 9781426216527 (hardcover : alk. paper)
LCSH: Astronomy--Juvenile literature. | Star Trek films--Juvenile literature. | Star Trek television programs--Juvenile literature. | Solar system--Juvenile literature. | Outer space--Juvenile literature.
LCC QB46 .F365 2016 | DDC 523--dc23
LC record available at https://urldefense.proofpoint.com/v2/url?u=http-3A__lccn.loc.gov_2016007552&d=CwIFAg&c=uw6TLu4hwhHdiGJOgwcWD4AjKQx6zvFcGEsbfiY9-El&r=5tkZtwhQ_OZM4AY3_OlYonuadShIJ0uA9xxRIR0Erks&m=5DNQVDNUiBHR7RGfsZuTLafa00C2gfNr39mwG39b9og&s=LUmSMyZDUaYywcpvNoZIAOeADRVcrybW_cOOjMGEvXA&e=

Interior design: Marty Ittner

Astronomy consultant: Beth Hufnagel

Printed in the United States of America

16/QGT-RRDML/1

THE IMAGES. THE STORIES. THE DISCOVERIES.

The books that bring them all together.

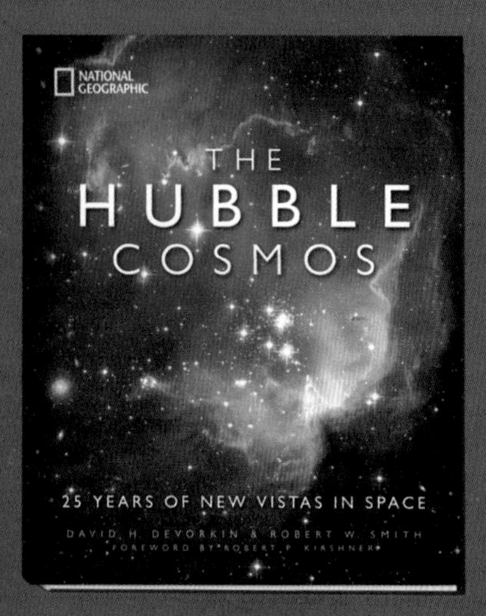

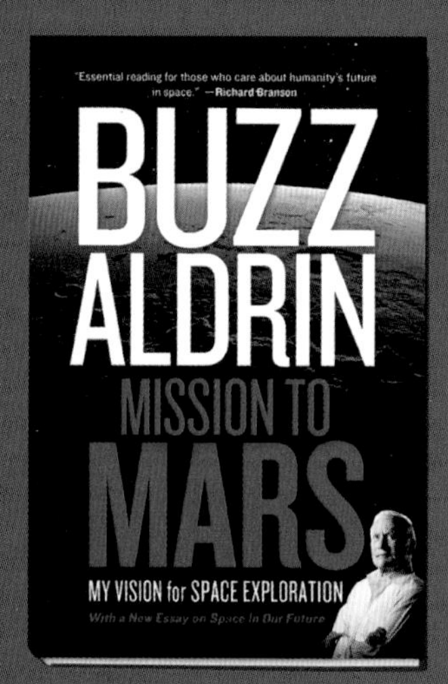

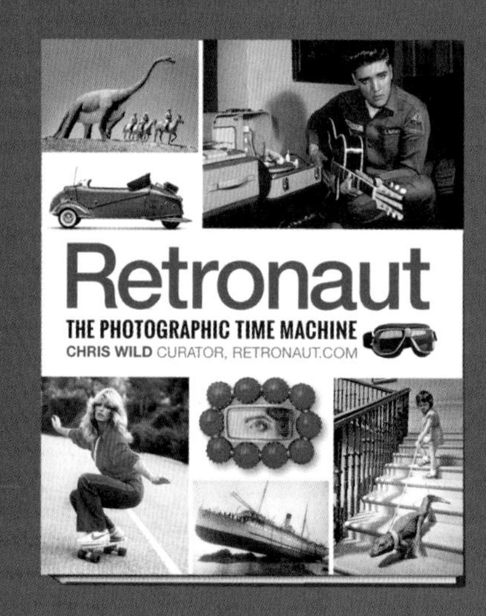

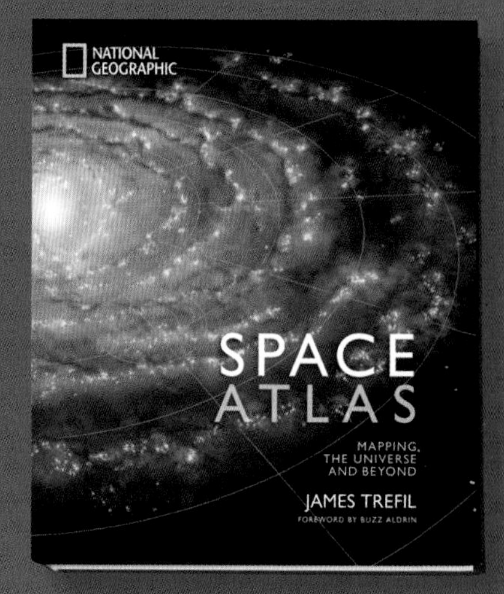

AVAILABLE WHEREVER BOOKS ARE SOLD

and at nationalgeographic.com/books